Deep Stuff

SCIENCE,

BIG BANGS,

MYSTERIES,

THE QUEST OF

TRUTH,

AND THE

VASTNESS

OF THE UNIVERSE

BY DAVID F. KOZIEL, JR.

Deep Stuff

ISBN-13: 978-1542530545

ISBN-10: 1542530547

This paperback edition was produced
in the United States of America.

Deep Stuff

IF YOU ARE PREPARED, you are invited to join along on a journey which solicits answers for some of life's deepest questions. DEEP STUFF is a scientific and philosophical treatment to existing concepts in physics, astronomy, history, and spirituality – one which explains pertinent issues in a comprehensible way, while making genuine contributions towards greater general understanding. Such topics explored within include the history of the universe, the Big Bang and the relentless expansion of space-time, dark matter and dark energy, black holes and event horizons, the birth and death of planet Earth, and possibilities for life elsewhere in the cosmos. Along the way, considered is the indescribable vastness of existence; and, how all of this relates to the human condition. From a fleck of dust in an ocean of eternity, this is the exploration of the depth of reality; a quest of enlightenment by humankind, pursing a feeble attempt to understand it all.

To my family and close friends,
for their support and guidance in all that I do,
no matter what that currently may be.

TABLE OF CONTENTS

Preface

T HE HEAVENS ARE IN A CONSTANT STATE OF FLUX; stars zoom about at hundreds of miles per second; galaxies speed to and fro faster still. Yet, the distances are so vast, and these celestial bodies so far away, that even in the entire span of a human life we see but a single, fixed cross-section.[1] We are very tiny – almost unimaginably tiny. Despite this, we are paradoxically driven by an insatiable curiosity to not merely *know*, but to *understand* this place we call the universe. We are but small creatures clinging to a planet for no particular reason other than it is where we fell. Together, we all circle the sun – which, in turn whirls around the Milky Way, following in the example of *billions* of other stars centered about a four-million solar-mass supermassive black hole.[2,3] Here is our home – a tiny, dusty, rocky speck, known affectionately as Earth. The humbling fact is, in this world of extremes, the scale of what we have thus far described remains still so incredibly insignificant compared to the rest of the cosmos, we barely possess words to describe it. These innumerable places; of countless galaxies, and the stupefying number of stars within; and, of the many planets which orbit them – there are not mere billions, not mere trillions, but far more than we have devised numbers to count.[4]

1   Turner, D. G. "An Eclectic View of our Milky Way Galaxy". (2013). Canadian Journal of Physics. (Vol. 92, No. 9).

2   Genzel, R.; *et. al.* "The Galactic Center Massive Black Hole and Nuclear Star Cluster". (2010). Reviews of Modern Physics. (Vol. 82).

3   Richstone, D.; *et. al.* "Supermassive Black Holes and the Evolution of Galaxies". (1998). Nature. (Vol. 395).

4   Bennett, J. O.; *et. al.* "The Cosmic Perspective: Fundamentals". (2009). Pearson.

I

Out of the Milky Way

The above image depicts the object known as *NGC 1300*, a so-called *"barred spiral"* galaxy, located approximately 69 million light years from Earth.[1] Though invisible to the unaided eye, it has approximately the same size, appearance, and structure as would be expected of our own galaxy, *The Milky Way*, if viewed from afar.[2,3] Image provided courtesy of NASA, ESA.

THE IRONY IS, that some of the greatest individuals who have changed the world, did so by looking away from it – for it was those insightful few who looked up before us and realized, that to gaze into the heavens, was to behold the mind of God. It was those who made the shrewd realization that looking through space – to islands of light in a sea of blackness – was also looking back through the ages, through all of history, upon all the times and places that ever were – and, destined to be. It was espying dioramas already unfolded unfathomably long ago, and in places so very far away.

Whether from learned awe or through blissful mystery, virtually all of us can appreciate the splendid wonder of the night sky. Dazzling, in its innumerable twinkling points of light – it is quite startlingly where the answers to some of our deepest, innermost questions are kept. Modern science has given us the surprising revelation that the heavens will oblige and answer us, most willingly and without censor; we need only figure out how to ask.

I continue to find it to be nothing less than astounding, that given the countless thousands of generations which preceded us – you and me – it has only been the last comparative few whom have been able to unravel even the most fundamental understandings of the heavens above. Comets composed of ice and dust – and not evil spirits; or, knowledge of what a star fundamentally is, and how it shines; nebulas, and supernova; and galaxies and their nearly incomprehensible size. All of these have been complete unknowns to the generations beyond our oldest photographs. One needs merely the fingers from one hand, to count the generations before who have discerned what it is that makes the sun glow;

or, that all the stars in the sky are other suns, and how very far away they are; or, even what a galaxy is – let alone that galaxies in the cosmos greatly outnumber stars within our own. All else who came before at best only have speculated using their wild imaginations, what each twinkling glimmer of light in the night sky could be.

Nicolaus Copernicus, born more than half a millennium ago in eastern Europe, was a lawyer-turned-astronomer who found himself annoyed and dissatisfied with the unscientific "astronomical traditions" of his day, feeling they inadequately explained the motions of the stars.[4] He was one of the first revolutionaries to champion modern astronomy, setting forth radical new ideas, placing the sun – not the Earth – as the central body which all of the planets revolve around. Unfortunately, such radical ideas were met with resistance by the then-contemporary, unerring self-proclaimed authorities on heavenly bodies (chiefly the clergy). As such, it was as late as the seventeenth century before astronomers began to speculate in earnest that the stars we observe in the night sky may in fact actually be distant bodies much like our own sun; yet, a reliable way of discerning *just* how far away the stars were – or simply proving this theory was even correct – remained elusive.[5] Without the proper tools, it proved to be tremendously difficult to validate such ideas as Copernican heliocentrism, or that each twinkling glimmer of light in the night sky was in fact other "suns". These ideas also challenged centuries of dogmatic religious ideas which attempted their own explanations regarding the motions of planets and stars.

In the following century, it was none other than the celebrated scientist Isaac Newton, who built upon the ideas of other great intellectuals who came before him. Through his endeavors, he gave to the world some of the most essential tools yet to the unraveling of the mysteries hidden in the sky above. One was by way of his advances on the subject of mathematics; *calculus* is indispensable to not just astronomy and cosmology, but virtually every branch of science. Other areas of specialization included optics, and the predecessor science of chemistry known as alchemy; as well as classical mechanics, and significant works on theology.[6]

Newton is widely regarded as the inventor of the first viable reflecting telescope.[7] The instruments used by the likes of Galileo and Johannes Kepler and their contemporaries suffered from a problem known as *chromatic aberration* – as all simple refracting telescopes do.[7,8] As light passes through a lens (just in the same manner as a glass prism) the various component colors have a tendency to become separated, and imprecisely focus. This manifests in what is described as a "rainbow halo" or "color fringing", which mars and degrades the resultant image. Additionally, as light passes through an optical lens, certain wavelengths are necessarily absorbed by the medium, potentially producing a false-color, or color-deficient image. Such collective distortion limits the capabilities of the telescope,

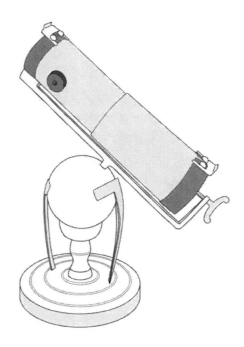

Sketch of Isaac Newton's reflecting telescope,
as first presented to the Royal Society (*circa* 1670).

effectively lowering optical resolution and obscuring image detail – thereby limiting scientific utility.

Newton's design circumvented the most vexing problems of refractive optics, dispensing with lenses in favor of mirrors. By reflecting light instead of bending it, primary sources of chromatic aberration were drastically reduced, facilitating the sharper observations necessary to advance general knowledge. Incidentally, as telescope design shifted to mirrors, it also dispensed with some of the lens' other difficulties. Foremost, a lens can only be made so large before it becomes prohibitively heavy under its own weight; additionally, due to both surfaces of the lens being optically active, the entire lens must be supported by only its edge.[9] This is problematic, as scientists require physically large optics to discern meaningful detail from far away targets; some of the largest telescopes today would be an impossibility of material science if a lens was the only viable option. Finally, since light must pass through the material body of a lens, the quality of resultant images are subject to optical flaws and imperfections.

A mirror, on the other hand, suffers from few of these issues which plague the transparent lens. Modern astronomers make use of large mirrors constructed in a variety of mediums, from transparent glass to opaque metals, to more exotic materials still.[10] Such optics can be made a fraction of the thickness and weight of their would-be equivalent refractive counterparts, and boast the capability to be supported from their entire obverse side rather than solely by their edges. Additionally, only the surface is optically active, limiting the degree to which material flaws degrade resultant images. All of these innovations synergistically allow the telescope to be built larger and to gather more light, increasing optical resolution, distilling ever finer detail from the cosmos. This, in the likeness of Newton's original design, has become the *de facto* standard for both present day ground-based and space-based telescope configurations. More recently, newer catadioptric arrangements have now become prevalent on most professional apparatuses, including many space-based observatories. Though these are significantly improved, their designs ultimately derive from the same overall elements of Newton's original precedent.[11]

Following its successful introduction, the optical reflecting telescope underwent rapid improvements during the eighteenth and nineteenth centuries. During this time, reflecting telescopes radically increased in size and capabilities, revealing ever finer details at progressively more extraordinary distances across space and time. All the while, these advances briskly paced our understanding and comprehension of the heavens. It became increasingly evident that the cosmos was tremendously vast, far beyond our wildest conceptions to date. Yet strangely, the more we probed, the more it seemed evident that these distant bodies followed rules which are no different than those observed on Earth. That is to say, the behaviors of planets and stars seemed to be dominated by the same rules which dominate our everyday lives. This realization represents perhaps one of the single most important observations that the human species has discerned from reality. Its occurrence marks the moment we have turned our gaze away, looking out of our galaxy, the Milky Way; it marks the moment we realized that all we had ever known was but an infinitesimal speck within a grand universe of possibility. If there is a place where the rules of nature truly are different, such a place exists increasingly far away (and long ago) as we consistently observe bodies behaving in identical, predictable manners gazing deeper and deeper into the darkness. To the extent we understand today, the existence of such differing rules of nature represents a statistical impossibility.

The most overt and familiar of these natural rules is of course gravity. It is the force that, for the majority of human history, has been associated with things clinging to the Earth – not explaining heavenly affairs. How can it be, that the same attraction between you and your chair as you read this book, could have definitive relevance to the stars, comets, or the Moon? As if enough were not already ascribed to him, it was none other than Isaac Newton who put forth his theory of *universal gravitation* in the year 1687 in his prodigiously influential masterpiece, *Principa*. He supposed, if it was by the same phenomenon of nature that the Moon "fell" around the Earth, just as an apple falls from a tree. However, unlike the apple, the Moon mysteriously never impacts the ground. Newton was able to (with varying degrees of accuracy in subsequent revisions) publish calculations which suggested that the Moon possessed a great horizontal, tangential velocity; in other words, the Moon was in *orbit* with the Earth.[12] Whether or not the falling object in question was indeed the infamous apple remains ambiguous; in any case, mankind once again inured to the benefit of Newton's fierce insight. The world would have to wait more than two-hundred years before someone came along who explained the world just a little better.

The attested oddity of space-time as described by Albert Einstein offers precisely such an explanation. On human scales, Newton's laws serve as a remarkably accurate estimation of reality – it is only when we investigate the very large, or the very small where discrepancies become apparent. Newtonian mechanics (or *"classical mechanics"* as it is often termed) is usually the most convenient framework for working out nearly every physical or engineering problem important to human activity; this convenience is largely the result of treating space-time as universally static. Only when relative speeds present a large disparity do considerations regarding the topography of space-time become relevant. This is not something which remarkably presents itself in daily life. In everyday occurrence, perhaps the most significant deviation from an ideal static representation of space-time that would be confronted by the general public is in air travel, yet the impact is so miniscule that it is imperceptible to all but our most sensitive equipment. Even space-faring astronauts, traveling at velocities more than an order of magnitude higher, experience trivial deviations from classical mechanics. Modern humans, in their common, general circumstances, simply do not have interactions with things moving at appreciable fractions of the speed of light – at least not until we direct our attention away to otherworldly fixations.

Unsurprisingly, this is where the *theory of relativity* acquires its name. Despite that this framework arguably represents the most complete and accurate description of the natural world to date, most people have little use for such extreme accuracy. Classical mechanics explains the world with much less effort, and in most situations, it is the most appropriate framework to use. It is only recently that we have created reasons to so precisely describe nature – at great expenditures of intellect. Mostly this has been done in application of technology to increase efficiencies; with the general intent to free our minds to more noble, if not more stimulating pursuits.

In contemporary times, improvements in telescope capability are increasingly supplanted by the application of advanced technology. Computing is the first of such technologies which often comes to mind, but equally advances in material science, optics, and mathematics contribute as well. Reciprocally, many of these technological advancements are either the direct or indirect result of applying something we have learned from the practice of observing the starry sky above. Furthermore, the realized benefits of such advances in space technologies are also frequently applied to benefit humanity in a general way.

A well-promulgated example of such technology adopted to inure to the benefit of the masses is the global positioning system, colloquially referred to as GPS. The accuracy of GPS is directly dependent upon knowledge derived from our understanding of distorted space-time – particularly, general relativity. Such understanding itself is a result of discoveries made via observation of heavenly bodies. Specifically, the application of relativity comes from correcting clock synchronization as a result of distorted space-time.[13,14] Such relativistic effects will cause a time disparity of 1.7µs to accumulate, *every day*. This is quite significant; in just this brief span of 1.7µs, an electromagnetic signal can travel more than half a kilometer. There is a generally held expectation that cellular phones and mobile devices should attain accuracy to within a few meters. Technologies used in aircraft navigation routinely achieve tolerances an order of magnitude higher. [15,16] This makes it comparatively easy to understand why correcting for such minute space-time distortions is so important, given our pervasive and ubiquitous reliance on these technologies. Applying a few lessons of observation in the vast cosmic distance has paradoxically given us this wisdom as to never become lost in our small world. Such is an example of our strange relationship with the all-knowledgeable sky.

Arguably the colloquial contemporary symbol of brain-power in general, Einstein himself struggled for many years teasing out the details to his renowned field equations of *general relativity*. First publishing his work in 1915, the theory formed the basis of which for the last century has, time after time on countless accounts, proved to be our most complete and accurate explanation of the physical world.[17] This came as an add-on to his previous *special relativity*, a groundbreaking theory in its own right. As evidenced by the decades-long progression of these subsequent theories, Einstein's epic mathematical crusade was rife with disappointing and frustrating difficulties; but a quest in vain, it was not.

Einstein's troubles working out the details of special and general relativity emphasize the reality, echoing that our current understanding of all existence has been anything but a clear-cut path; in fact it has never even been a singular path. The network of roads to ultimate understanding has and shall forever be littered with innumerable detours and dead ends. The frontier of knowledge, as it always has been, is no less murky now than it ever was. One may suppose, that this is the nature of progress. We must be always be willing to accept change, provided there is sufficient justification, sufficient evidence. It is not just astrophysics, or physics in general – all of science over the ages has been riddled and fraught by these unfortunate treacheries. Trepanning, bloodletting, witch-hunting, human sacrifices, superstitions – the list of bad ideas go on.

Yet somewhere, somehow, we inevitably get it *right*. What once appeared murky, eventually becomes crystal-clear; our galaxy – our home in the cosmos – is aptly named to reflect this fact. In these modern times, our vision has transcended the Milky Way; we have seen so much farther than humans ever have before. We are now finding ourselves gazing into the abyss at countless milky islands which belong to not us, not to our Milky Way, but are whole and distinct galaxies in their own rite. There is a certain irony here, in that from moment to moment, with the more we learn, the more we can look in retrospect upon ourselves, and, marvel at how little we knew back then.

1. Sheth, K. "Barred Spiral Galaxies and Galactic Evolution". (2008). National Aeronautics and Space Administration (NASA), Goddard Space Flight Center, and the Space Telescope Science Institute.

2. Aguerri, J. A. L.; *et. al.* "Characterizing bar structures: application to NGC 1300, NGC 7479, and NGC 7723". (2000). Astronomy and Astrophysics. (Issue 361, 841 – 849).

3. Atkinson, J. W.; *et. al.* "Supermassive black hold mass measurements for NGC 1300 and NGC 2748 based on HST emission-line gas kinematics". (2005). Monthly Notices of the Royal Astronomical Society. (Vol. 359, 504 – 520).

4. Contopoulos, G. "The Astronomy and Cosmology of Copernicus". (1974). Highlights in Astronomy of the International Astronomical Union. (Vol. 3, 67 – 85).

5. Danielson, D. "The Case Against Copernicus". (January 2014). Scientific American. (Vol. 310, Issue 1, 72 – 77).

6. Rickey, V. F. "Isaac Newton: Man, Myth, and Mathematics". (1994). College Mathematics Journal. (Vol. 18, 362 – 389).

7. "Reflecting Telescopes". (2001). Encyclopedia of Astronomy and Astrophysics. Nature Publishing Group.

8. Westfall, R. S. "Galileo and the Telescope". (1985). Science and Patronage. (Vol. 76, No. 1, 11 – 30).

9. Onah, C. I.; Ogudo, C. M. "Design and construction of a refracting telescope." (2014). International Journal of Astrophysics and Space Science. (Vol. 2, No. 4, 56 – 65).

10   Luz, P. L.; Rice, T. "Mirror Material Properties Compiled for Preliminary Design of the Next Generation Space Telescope". (1998). National Aeronautics and Space Administration (NASA), Marshal Space Flight Center.

11   Simpson, D.G. "Optics of the Hubble Space Telescope". (2010). Prince George's Community College, Department of Physical Sciences and Engineering.

12   Kollerstrom, N. "Newton's two 'Moon-tests'". (1991). British Journal for History of Science. (Vol. 24, 369 – 372).

13   "Global Positioning System Standard Positioning Service Performance Standard". 4th Edition. (2008). United States Department of Defense.

14   Ashby, N. "Relativity and the Global Positioning System". (May 2002). Physics Today. (Vol. 55, Issue 5, 41 – 47).

15   "Guide for Ground Based Augmentation System Implementation". (2013). International Civil Aviation Organization (ICAO).

16   "Instrument Flying Handbook". (2012). The United States Department of Transportation, Federal Aviation Administration (FAA).

17   Kox, A. J.; Klein, M. J.; Schulmann, R. "The Collected Papers of Albert Einstein". (1997). Princeton University Press.

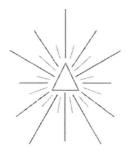

II

TRUTH

"I do not know what I may appear to the world; but to myself I seem to have only been like a boy, playing on the seashore, and diverting myself, in now and then finding a smoother pebble or a prettier shell than ordinary, while the great ocean of truth lay all undiscovered to me." [1]

– Isaac Newton,
remarking near the end of his life.

HISTORICALLY, the concept of *trivium* was taught since antiquity as the foundation principles necessary in the framework for meaningful questioning of the world. It was postulated that through the trinity of grammar, dialect, and rhetoric, mankind gains the ability to interpret the world with certainty – a convening on *truth*. From this foundation, it was reasoned to be possible to explore the "arts", then held to be the study of numbers, geometry, music, and astronomy – otherwise referred to as the *quadrivium*.[2] Today, we place more weight on rationalism, reason, and deduction than theology and empiricism. Over the last few hundred years, through our collective and collaborative advances in the reflective analysis of thought and reason, we have deduced a more fundamental understanding of what it means to be "certain" – that which is the elementary state of *truth*. As a result, our divisions of academic study have been refined accordingly, explaining the world which we inhabit with an unprecedented degree of accuracy and insight.

When man is hungry, he seeks food; when cold, he seeks the warmth of fire. When he is chased by a snarling, charging lion, he runs, and does so briskly. But when man is satiated, that is when in abundance of the necessities, he does what no other creature does. He thinks; he ponders the world – and suddenly, he is no longer content. He has abruptly become dissatisfied with something – but what, specifically? Perhaps, an incongruity or omission in his world – he has noticed something which he cannot make sense of, something which cannot be explained with his usual, customary explanations. Yet he knows how to feed himself, if he again becomes hungry; he knows how to warm himself if he becomes cold. He knows *exactly* what to do when he sees that angry lion again. All the same, he remains bothered by his newly found, incompatible notion. But *why*, we must ask. To what end is this seemingly perverse insistence on perturbing a completely satiated state of well-being?

Such innate compulsion is a fundamental consequence of our relationship with the concept of truth. Instinctively, before we have knowledge of the rigors of science, our minds are already hardwired to demand evidence to explain reality. At any given moment, we possess an arbitrary amount of evidence validating the conditions of the present. However, to an extent, we realize some sources are dubious unless they are logically evaluated. This is to say, we are more apt to accept evidence which was acquired through logical (and hence reproducible) means. One could say the more evidence we acquire, and the better such evidence is logically evaluated, the more confident we are in believing something. When that belief is something which holds priority – such as a matter of life or death – we want to be as certain as possible that we have not erred in our evaluation. Accordingly, such an iterative "loop" runs within our consciousness, incessantly and perpetually evaluating the present; it continually searches for the weakest logical support of truth in evidence, and though efficient use of our idle time, and repeatedly forces us to reconcile our explanations.

Human beings are not *physically* well adapted for survival, outside of the infrastructure of society. Armadillos have thick skin; vipers have a venous bite; roses have sharp thorns; and so-on. The snarling lion is not always concerned about being correct, because it does not need to be. For us, our intellect is perhaps our greatest survival advantage – thus it is no wonder that we so obsess over what is true and what is false.[3] Likewise, we have many unpleasant feelings – betrayal, anger, sadness, among others – when we are deceived or deprived of the truth. This *truth*, or state of truthfulness, is an ingrained idea – an obsession so deeply coveted by consciousness that it becomes its defining characteristic; so much so, when removed, human consciousness ceases to be recognizable as contiguous or sane. Not only is it a particularly important concept, it is perhaps the most fundamental postulate of all time – that which is the meaning of *truth*. It is the foundation principle of all conscious thought; the common denominator to which all other ideas are measured.

The philosophical examination of truth is by no means anything new. Throughout the ages, there have been long and tortuously drawn out works rigorously examining this dichotomy of true versus false, attempting to define what the essence of truth really is. For each theory, there can be – and certainly have been – entire books ascribed to their explanation; but I feel as though the definition of truth should be more basic, more essential. After all, it is *the* fundamental idea central to all others. The answer should be a simple one.[4]

To many, the identity of truth is simply what "*is*" – in other words, that which is infallibly observed. Aristotle, a forefather of philosophy, suggests definitions for true and false, oft-quoted as:

> "*To say what is that it is not, or what is not that it is,*
> *is false, while to say of what is that it is and what is*
> *not that it is not true.*" [5]

But how do we know when our senses deceive us? In other words, how are we to differentiate between what is actually true, and what we happen to think is true – that which merely *appears* as true? Perhaps we can say truth is that which is not a lie; but this attempt merely inverts the logic – it is the same idea stated another way. Furthermore, the concept of a lie usually insinuates deception. It does not *define* truth.

When we seek to define something, the logical place to seek such definition is, of course, the dictionary. Yet practically universally, the essential definition is somehow evaded – in its place we are met with vague and often recursive explanations of this essential thought. Even the best dictionary definitions which could be found, unsatisfyingly state truth is "*in accordance with the actual state of affairs*" or, "*that which is true or in accordance with fact or reality*".[6,7] These are interesting associations and observations – but nonetheless unconvincing that any would be the most fundamental definition.

So, the question persists; and still, we ask, what is *really* meant if something is actually true – or untrue? This demands further rhetorical analysis.[8] We have already suggested that a body of evidence which has been obtained through logical means points to the truth. However, the question remains as to precisely where the distinction is made – between something which *might* be true, versus something which is *actually* true. Without elaborating further, it appears we have left truth to be interpreted in the eye of the beholder. Perhaps, in order to reconcile, we can reason that truth is the belief that the greatest number of people *consider* to be true – in other words, what the majority believes. But this notion fails for subjects such as mathematics, where everyone in a particular sect may believe that one plus one equals three, until that one person proves the answer is in fact *two*. Mere beliefs are irrelevant in terms of absolute, essential truth; our trust in something does not guarantee its infallibility. Truth must be something more definitive, and less dependent on our delicate feelings and emotions.

There is also the consideration that perhaps truth is something based on *intent*, and likewise the purest intents correlate most closely with the purest form of the truth. In a sense, this meaning consequently implies that we find only what we are looking for. Although I can agree this may be the case more times than it should be, I also hold the belief – even if such beliefs are ultimately irrelevant – that the purists' pursuit of truth shall staunchly reject such pitfalls. Beliefs aside, if we rhetorically analyze the former, this line of reason suggests that truth is based on the genuineness of the subscriber – but we are quite familiar with the unfortunate circumstance that even the most genuine intentions may be, at times, misguided. No matter how good the motivations, however pure the intent, one plus one still must *always* equal two, and not sometimes three. Otherwise, disaster would ensue as truth failed and reality imploded upon itself.

Therefore the best definition that I can independently come up with, is that essential truth is *the infallible notion that the world operates in a consistent manner* – that is, every identical situation under identical states and conditions will always yield identical results. Truth describes a framework of reproducibility which guarantees the world around us never finds itself in a paradox. To personify human behaviors, the universe will never tell lies. Such may be one of the most important concepts of all time.

Alas, despite the importance of this notion, it by no means is the end-all, be-all retort to the matter at hand. Insomuch that we have established the *nature* of truth, we have yet to outline a means for proving its *identity*. We have already established that the direction of truth is likely indicated by logic and logically obtained evidence.

By extension, we may require truth to initially be described as a thought which is logically sound – something which now requires us to define what logic is, in order to equate it to truth. Logic is the study of reasoning, and identifying *what* correct reasoning is; logic therefore reveals truth by separating out states of being false. It is a particular branch of study based on consistency and reproducibility; it underscores the importance of a universe which behaves in a predictable and reproducible manner – that is, never leading to a situation which results in a paradox. However, just because something is logical, does not necessarily mean it has happened, or will happen. Logic identifies truth in part, but not unequivocally.

Is evidence not the classical legal criteria for weighing in on the truth? While logic is regarded as abstract, evidence is very real; it becomes the physical manifestation of the conditions and outcomes of a dialectic – but unto this, logic and evidence must be inseparably bound. Our innate association between the two is rhetorically robust, a pillar upholding the truth. But, like its legal counterpart, we also have to look at how this evidence was obtained. Evidence which was obtained illegally is not permissible in a court of law; likewise, evidence suggestive of a particular truth which was obtained *illogically* must also be dismissed, ignored.

Thus far, we have elaborated on this essential description of truth to define it as the intersection of logic and evidence. If, however, logic validates evidence, and likewise supporting evidence validates logic, there exists a critical flaw in our theory. How would it be known if our logic was sound yet the evidence was not; or *vice versa*? The framework, as presented, lacks a system of checks and balances; the logic–evidence basis is recursive. There must be some additional criteria which serves to validate each of logic and evidence, unto which the two also validate it.

Often times we regard mathematics as the language of the universe; but it was not our invention. We make use of pencils and paper, constructs of abstract ideas, calculators and computers – just in the same manner as early humans' control of fire was via sticks and torches and other various implements. The value and power of mathematics stems from its universal utility. It is a description of the predictable and consistent way in which the world around us works – in an abstracted sense, the patterns which emerge out of the rules which the universe contracts to obey. Something becoming increasingly suggestive in modern times through the study of quantum mechanics and theoretical physics, is the notion that mathematics is one manifestation of nature's unrelenting obsession with quanta.

To humans, in the most rudimentary sense, the field of *mathematics* can be described as the study of numbers, and their relationships thereof. We must remember that numbers are an abstract construct; they are of our own invention, the intellectual animals of the Earth; a tool. We regard this study of numbers as a particularly important contribution to the basis of reality and truth, in that it describes the universal consistencies provided by the cosmos. What we are beginning to describe, is the *Evidence-Logic-Mathematics axis* – the interdependent relationships of reason which function in unison to identify truth.

To elaborate further, evidence put forth must be both logical, as well as validated by mathematics; in turn, mathematical results must also agree with the presented evidence; furthermore, the validity of these mathematics depends upon the rules of logic. Finally, logic must support the evidence and the manner in which it was obtained, and also be in agreement with the mathematical calculations which describe the state of the evidence obtained. The region where all three overlap – that is, a convergence of logic, evidence, and mathematics – is what we know as *truth*.

The application of scientific thinking to explain natural phenomena represents one of the most significant advancements in humanity's quest for ultimate understanding. Science yields evidence; however, it is important to note that evidence alone does not by necessity yield truth. To that end, some form of mathematics is employed – most often statistics – to reconcile logic and evidence; it is only by agreement of such which serves to bridge our scientific discoveries into the domain of truth.[9] Of logic, mathematics, and evidence, each discipline can be thought as to form a "basis attribute", tracing the derivation of truth. Their

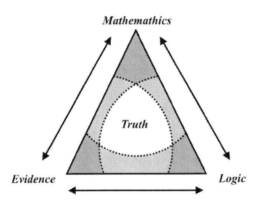

The Evidence—Logic—Mathematics Axis

interdependence ensures that no one component precedes another in importance – all are required to agree in order for *truth* to exist. Likewise, failure of any of these three pillars violates truth, and thus describes a state of falsehood.

Statistics is one branch of mathematics which sees itself as perhaps the most frequent exercise of this "bridge" between logic and evidence. Consequently, it also represents the most practical and accessible application. We may refer to such a determination as *"truth in probability"*. Essentially, it implies that if one is going to make an assumption or a recommendation based on a given probability, then one must also accept and treat all other equal probabilities in the same manner. Otherwise, biased decisions are made based on emotion or past experiences, and lack a definitive basis in truth or reality.

It may be helpful to consider an example. If one postulate has only the likelihood of a one-in-a-hundred chance of being correct, and we accept it as false, then we must also accept all other scenarios which are one-in-a-hundred chances as unconditionally false. The more we favor one or the other, the more proportionally skewed the probabilities will become. If we are twice as likely to accept one equal probability scenario as just described above, over another equivalent scenario as truth, then the odds are no longer both one-in-a-hundred, but rather one in approximately sixty-seven, and one in approximately one-hundred thirty-three. Additionally, if we favor one possibility over the other, in the case where we accept that both possibilities have equal *probability* of being true, it becomes that we have necessarily renounced our belief that the universe behaves in a consistent manner. One plus one now sometimes equals something other than two; hypothetically, we may be arguing that the result of one plus one varies somewhere between one and three. Mathematics fails, and consequently our argument becomes *illogical* – there is no convening on truth. Ergo, all truth becomes meaningless.

Though statistics may represent the most common means to satisfy the mathematical validation, it does not constitute the only. Many other specialized applications of mathematics are able to transcend simple probabilities, as we relegate to the abstract far more complicated patterns of nature, via such constructs as fractions, ratios, statistics, algebra, trigonometry, geometry, calculus and so-on. The logic may also become somewhat more complicated, as does the body of evidence collected and thus the associations made. However, all of this is remains merely a more elaborate implementation of this simple idea, the *Evidence-Logic-Mathematics axis*. In conjunction with the scientific method, such may be our best tool thus far for distilling truth from our senses and the thoughts in our minds.

1   Mandelbrote, S. "Footprints of the Lion: Isaac Newton at Work". (2002). Cambridge University Library. University Press, Cambridge.

2   Gossin, P. "Rhetoric". Encyclopedia of Literature and Science. (2002). Greenwood Publishing Group. (386).

3   Kaplan, J. M. "Historical Evidence and Human Adaptations". (2002). Philosophy of Science. (Vol. 69, 294 – 304).

4   Davidson, D. "The Folly of Trying to Define Truth". (1996). The Journal of Philosophy. (Vol. 93, Issue 6).

5   Ramsey, F. P. "Truth and Probability". (1926). Foundations: Essays in Philosophy, Logic, Mathematics and Economics. (1931, 1978). Humanities Press. (58 – 100).

6   "Truth". Merrian-Webster.com. (*n.d.*). Retrieved August 2014.

7   "Truth". Oxford English Dictionary, Oxforddictionaries.com. (*n.d.*). Oxford University Press. Retrieved August 2014.

8   Tarski, A. "The Semantic Conception of Truth". (1944). Philosophy and Phenomenological Research. (Vol. 4, No. 3, 341 – 376).

9   Popper, K. "The Logic of Scientific Discovery". (1935, 2005). Routledge Classics, London and New York. (2002). Taylor & Francis Group. (2005).

– III –

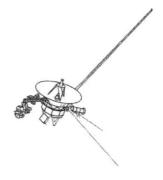

INTO THE DARKNESS

"Science is an attempt, largely successful, to understand the world, to get a grip on things, to get hold of ourselves, to steer a safe course. Microbiology and meteorology now explain what only a few centuries ago was considered sufficient cause to burn women to death." [1]

– Carl E. Sagan

SPACE IS A DARK, LONELY PLACE. It is cold, uninviting, and extraordinarily hostile to all known forms of life. Death can come of any number of a thousand ways; survival is not guaranteed; it is merely an accepted probability. Yet, for some yearning, unexplainable reason, we are perplexingly almost universally compelled to know its secrets, to explore its wonders. Despite the perilous dangers, the costs, the challenges – the countless reasons to the contrary – we are inexplicably attracted to the darkness. Perhaps it may be attributed to general curiosity, or the innate instinct of humankind to explore what is unknown. It is equally peculiar that our endeavors have been so greatly rewarded. Discoveries made in the great dark expanse above have relentlessly proven enlightening to our comparatively mundane lives down below. Almost without exception, the religions of our world speak of the heavens and gods possessing omnipotence and of ultimate knowledge. How strange indeed, that this carries so much truth.

Today, there are no longer any previously unadventured ocean crossings; there are no continents yet waiting to be discovered. Every square mile of our planet has been charted and accounted for. In consideration of all of our history, this is a trying first for humans. The notion that beyond the horizon lies an endless expanse of nature is one which mankind will never again experience, on Earth. It is almost as if a fundamental freedom has been taken away, exhausted by the masses – there is no longer enough of our planet for everyone. Perhaps this is one factor which has suddenly turned our attention upward and outward. In a nautical analogy, one might suppose, deep black has become the new deep blue.

For the vast majority of human history, our ancestors have wildly speculated the nature of the stars, many times blindly contriving significance and worldly causes to their motions. Celestial events, such as the occurrence of eclipses, passage of nearby comets, and sudden appearances of brilliant meteors have conjured both imagined legends of boyish fantasy, and stricken panic in the minds of grown men. Fanciful names and tales have been ascribed to the stars which appear to return year after year with divine precision, as well as the "wandering" stars we now know as planets. Such perfection could only be the doing of gods, so we naïvely reasoned. If only the world had received the telescope sooner, perhaps thousands of years of conjecture may have been put to good astronomical use.

Technological advances, a better understanding of physics, and experimentation in rocketry in the decades prior – including the well-popularized research of Robert Goddard – all helped to lay the groundwork to make future endeavors possible.[2] Near the end of the second world war, however, our fascination with propelling things into the night sky took a dramatic turn. At the forefront of aerospace engineering was the German V-2 rocket, the first manmade device designed to function above the atmosphere in "space" – although this feat was secondary to its grisly primary objective.[3,4] The V-2, in terms of accessibility to space, could at best only be configured as a suborbital[†] vehicle; this meant it spent only a brief period of time in space on an arcing trajectory, which quickly returned it to the ground. Yet, it was not long before a more remarkable achievement was attained by the Soviets, when a few short years later they captured the unwavering attention of the world by lofting a beach ball sized metallic sphere into *orbit*. Just as Newton supposed the Moon perpetually "fell" around the Earth, so now did the gleaming satellite *Sputnik*.[5] Several more incarnations of *Sputnik* followed, as did American satellites shortly thereafter.[6] Though almost infinitesimally fainter, by the standards of ages gone by humans had finally taken their place amongst the gods.

The "space race" as it came to be known, was driven by fierce nationalist competition – in the beginning, chiefly between America and the Soviet Union. Accordingly, it was not long before tiny metal spheres were replaced by much larger ships containing men. On the fortuitous day of April 12th, 1961, Soviet pilot Yuri Gagarin aboard *Vostok 1* became planet Earth's first space-faring astronaut.[7] Others quickly followed, both from the Soviet Union and the United States. As these brave pioneers cruised around the entirety of the planet in mere minutes, they looked down upon everyone, seeing over all of the world. At first they carried only a single person, but before long they were carrying two, then three. Some years later, the space shuttle could

---

[†] The *Kármán line*, at 100 kilometers, represents the most widely recognized international altitude threshold for space.

carry as many as eight at a time. Orbital outposts were constructed – akin to tiny cities in the sky which perpetually circled the Earth. We even paid visit to some of the heavenly bodies, sending first robotic messengers ahead of our aspirations for manned expeditions. [8,9]

Le Voyage dans la Lune, or as translated to English, *"a Trip to the Moon"* is a French silent motion picture which debuted in 1902, approximately a year before humans had first taken to the sky in powered flight. [10,11] The film depicts a number of individuals embarking on a voyage to the Moon, whereby their capsule receives a cannon-assisted launch at tremendous speed, subsequently utilizing that same capsule to return safely to the Earth. The production takes obvious liberties in poking fun at the ridiculousness of such a feat – a conveyance of buffoonery for such lofty ambitions. In the ensuing years, in contrast to the portrayal in the film, it became increasingly obvious that the Moon was a barren wasteland, home to no extraterrestrial inhabitants, neither friendly nor unfriendly. However, it is still mind boggling to think a mere sixty-seven years later, three brave men sitting in a similarly sized and shaped capsule, were shot towards the Moon, faster than a bullet; and they utilized that same capsule to safely to the Earth, splashing down in the ocean, just as depicted in the film. Admittedly, science fiction can be strange; but science fact can often times be stranger.

Owing to our own biological limitations, and the unfriendliness of space, humans have thus far not yet ventured beyond the Moon – at least not in person. Robotic exploration has momentarily taken precedence. Presently, our innovations need time to catch up with our ambitions; our continued journey requires better technology. In lieu of this, we have travelled to all of the known major planets, and many of the minor ones. We have even landed on a few; this includes not only our own Moon, but the world known as Titan, the largest moon of Saturn. In our endeavors, we have found that clouds and rain, rivers and lakes, lightning and volcanoes all are not unique to this world. In recent years, we have gained the ability to forecast Earth's weather on a global scale, with increasing accuracy as well as progressively further outlooks. To that end, we have also begun to forecast the weather of Venus, Mars, and to an extent, even the gas giants. Solar-observing satellites fly in orbits around the sun, and forecast the extra-planetary weather of the solar system itself – chiefly such phenomenon as the solar wind, coronal mass ejections, and the eleven-year solar magnetic cycle.[12] In the future, the possibilities of new, yet-unimagined capabilities are uncertain; however, it is safe to say they will undoubtedly mesmerize us.

For now, our present achievements seem impressive enough. In the late summer of 1977 – in unquestionably the most audacious undertaking mankind has ever conceived – the United States launched the twin Voyager spacecraft on a fantastic trajectory towards the stars.[13] The primary objective of the mission was to facilitate the study of the giants of our solar system – Jupiter, Saturn, Neptune, and Uranus, and their planetary neighborhoods. In the ensuing decades, the *Voyager* program has given us some of the most breathtaking views of these worlds, and immensely increased our understanding of the solar system, including our place within it. Neptune, the mission's last planetary target, was observed by *Voyager 2* in a close fly-by during the summer of 1989.[14]

Since that time, the twin probes have left the solar system entirely, sailing on, never to return again. *Voyager 1* has become the most distant man-made object ever jettisoned from Earth, bearing the distinction as the first artifact of humanity to enter interstellar space.[15] The missions of these spacecraft still continue to this day, their trajectories which have taken them on a journey not just to explore the solar system, but to continue further.[16] Our understanding of the cosmos has been likewise improved in unexpected ways. In one example within recent years, the probes have mapped how the solar winds interact with the galactic interstellar medium. The *Voyager* probes are expected to continue to return data for some time, as they wander deeper into the unknown of interstellar space.[17] The most recent spacecraft placed on an interstellar trajectory was *New Horizons*, aimed at conducting science on the dwarf planet Pluto and potentially other Kuiper belt objects.[18] The now silent *Pioneer 10* and *Pioneer 11* probes are also on trajectories towards other stars.[19] Each of these missions has expanded our knowledge of reality in innumerable ways, many times revealing discoveries which were unexpected. Never before have we dared to ask the heavens for answers more zealously.

Consider the magnitude of our current ambitions for a moment. Not even one hundred years prior, those famous pioneering brothers Orville and Wilbur Wright were but children – mere grade schoolers. No airplanes had yet taken to the sky. Mankind in fact had not yet managed to fly *anything* heavier than air. True flight was but a fanciful dream. Within the limits of a single human lifetime – just several brief decades – we had become first pioneers of aviation, then daring astronauts; to auspicious interplanetary explorers; and now finally, a fledgling interstellar species. Most impressively, this is only the beginning. Into the darkness we shall forge ahead, towards the answers kept in the mysterious abyss. In the precedent manner that so many religious prophecies have presumptuously proclaimed – nearly since the dawn of recorded history – mankind will ascend into the heavens, and thereby achieve enlightenment.

1 Sagan, C. E. "Demon-Haunted World: Science as a Candle in the Dark". (1995, 2011). Random House Publishing Group.

2 Page, B. R. "The Rocket Experiments of Robert H. Goddard, 1911 – 1930". (1991). The Physics Teacher. (Vol. 29, No. 8, 490 – 496).

3 Quinn, M. C. "Historic American Engineering Record" (HAER). (No. NM-1B). (1986). United States Department of the Interior, National Park Service.

4 Tillman, B. "Germany's V-2 Rocket: A Lethal First Step Into Space". (October 2015). Flight Journal.

5 DuBridge, L. A. "The Challenge of Sputnik". (1958). Engineering and Science. (Vol. XXI).

6 Ludwig., G. H. "The First Explorer Satellites". (2004). University of Iowa.

7 "United Nations Human Space Technology Initiative (HSTI)". (2013). International Academy of Astronautics, 8th IAA Symposium on the Future of Space.

8 "Information Summaries: The Early Years: Mercury to Apollo-Soyuz". (1991). National Aeronautics and Space Administration (NASA), Kennedy Space Center.

9 "Space Shuttle Era Facts". (2011). National Aeronautics and Space Administration (NASA), Kennedy Space Center.

10 Méliès, G (Director). *Le Voyage dans la Lune*. Motion picture. (1902). France: Star Film Company.

11 "The First Flight". (*n.d.*). United States Department of the Interior, National Park Service.

12   Steenburgh, R. A.; *et. al.* "From Predicting Solar Activity to Forecasting Space Weather: Practical Examples of Research-to-Operations and Operations-to-Research". (2014). Solar Physics. (Vol. 289, No. 2, 675 – 690).

13   "Voyager to the Outer Planets and Into Interstellar Space". National Aeronautics and Space Administration (NASA), Jet Propulsion Laboratory.

14   Angor, C.; *et. al.* "The Exploration of Neptune and Triton". (2009). National Research Council (NRC) 2009 Planetary Science Decadal Survey.

15   "Jet Propulsion Laboratory". Fact sheet. (2016). National Aeronautics and Space Administration (NASA), Jet Propulsion Laboratory.

16   Stone, E. C.; *et. al.* "Voyager Interstellar Mission". (2005). National Aeronautics and Space Administration (NASA), Jet Propulsion Laboratory.

17   "Voyager: Spacecraft Lifetime". (2015). National Aeronautics and Space Administration (NASA), Jet Propulsion Laboratory.

18   Porter, S. B.; *et. al.* "The First High-Phase Observations of a KBO: *New Horizons* Imaging of (15810) 1994 JR_1 from the Kuiper Belt". (2016). The Astrophysical Journal Letters. (Vol. 828, No. 2).

19   Anderson, J. D.; *et. al.* "Study of the anomalous acceleration of Pioneer 10 and 11". (2005). Physical Review D. (Vol. 65, No. 8).

– IV –

VASTNESS

IN CEASELESS AWE, echoing in the fashion of the thousands upon thousands of generations of our ancestors who preceded us, upwards we stare into the starry night sky; we are entranced by the countless incredibly tiny shimmering glints of light – most of them smaller than the period at the end of this sentence. Curiously, in our gazing, we might choose a single, typical star and compare it to our own sun; such scale alone is testament to the enormity of merely our local stellar neighborhood.[1] Each point of light that we are able to resolve in the sky with our unaided eye is a star within our own galaxy; and, from our observations of the universe thus far, we appear to be in a *typical* galaxy. The number of stars within the Milky Way is vast; without the magnification afforded by a pair of binoculars or a telescope, we can only see an estimated few thousand or so stars on a given clear night; this is true even under the darkest and most ideal of conditions. The rest simply blend into the soft glowing fog known as the *Milky Way*. For however spectacular the view of the night sky seems, ponder this fact the next time you tilt your head in awe – you are only seeing, at the absolute most, 0.000005% of all the stars in our galaxy. This is an incredibly tiny number in comparison to current best estimates placing the number of stars within the Milky Way at between 100 and 400 *billion*.[2] In an attempt to put this disparity into perspective, consider that there are multiple orders of magnitude more stars in just our own Milky Way than there are people on planet Earth. And, as if this was not inconceivable enough, we must remember that we have not considered a single star within any "nearby" galaxies, nor any in the other hundreds of billions of other galaxies across the cosmos.

We are extraordinarily far away from our neighbors. It is very easy to look up and forget each and every one of those pin-point dots of light is actually an entire star; many of them far mightier than our own sun – and each the center of attention to another solar system, with many (if not most) having their own planets. Our own star, being somewhat average, actually pales in comparison to many of these other distant fireballs – yet they are so incredibly far away that their size is not even something we can distinguish in a relative way. Take, for instance, the supermassive star known as *Betelgeuse* within the constellation of Orion, *the hunter*.[3] It typically is visible year-round due to its proximity to the celestial equator, and is also one of the more prominent constellations in the night sky.

The main body of the hunter visualized by means of the sketch of the constellation on a following page, with Betelgeuse forming his left shoulder. I always find this fact to be impressive; Betelgeuse is one of the largest known stars in our galaxy, exhibiting a radius of some approximate 800 and 1,200 times that of the sun.[4] Each night, it is as visible as the Moon or the sun or any other common site; yet it is this solitary object of unimaginable proportions, and by far the largest single object we can see with the unaided eye. In a side-by-side comparison of scales, let us consider:

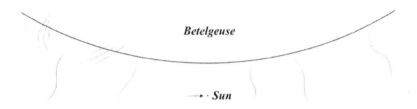

The arc depicted here would have to be extended full-circle to depict the entire star; this entire book would fit inside. Furthermore, no justice is given to the enormity of even our own sun – the tiny dot above which is smaller than the period at the end of this sentence – is large enough to contain over a *million* Earths. Betelgeuse, this pinprick of light, as seen from the vantage point of

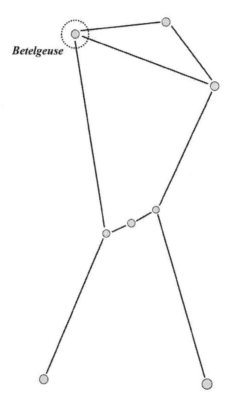

Main body of "the hunter" within the constellation Orion.
The star *Betelgeuse* is circled, upper left.[3]

the Earth, appears to wax and wane in intensity – a feature of the star's irregular, variable nature. Its outer layers are likely nebulous and unstable, exhibiting variable mass ejections as it teeters dangerously close to exploding in a spectacular supernova.[5] It should provide a spectacular light show sometime within the next hundred thousand years or so, if humanity is still around to witness it.[6] We consider Betelgeuse to be relatively close to us (at a little over six-hundred light years away) with the Milky Way galaxy itself being approximately one-hundred thousand light years in diameter. Yet, all we have is a point-like glint of shining, twinkling light, which compared to the thousands upon thousands of other stars, this colossal stellar monstrosity divulges little-to-no evidence to the unaided human eye of its gargantuan nature. Though it is relatively bright, brightness alone is not a good measure of size; there are stars which are much smaller than Betelgeuse that appear brighter from our vantage point. Scientists have to employ much more creative methods in order to determine the distances of stars from the Earth.

It is in fact some of the objects we casually refer to as the "smallest", perhaps a major asteroid such as Ceres (considered to be a dwarf plant, and the largest object within the asteroid belt), are actually much bigger than you may think. Even as a meager asteroid, it still has almost three million square kilometers of surface real estate – which is nearly twice the land area of the state of Alaska, by far the largest state within the United

States.[7] Even if you could come to grips with the enormity of celestial objects, there is also the consideration of the even more fantastic distances between them. Many people grossly underestimate the distances involved in circumnavigating our own tiny little crumb of a planet. In terms of everyday scales, the oceans are utterly huge, expansive spaces; however, more than a few have come to terms with this, and have circumnavigated the globe, first by sea. At the equator, the circumference of the Earth is approximately 24,900 miles.[8] Ships take *months* to round the world, and are perhaps best on par with the speeds we experience in everyday life. With the advent of powered flight, we can make the circumnavigation in a matter of days. Subsonic planes can cut around-the-world travel to around three days using a purpose-built jet aircraft.[9] The famed Concorde, flying at up to twice the speed of sound, once did a westbound circumnavigation in approximately 32 hours, including stops to refuel.[10] Spacecraft in low Earth orbit, which move at incredible speeds of tens of miles per second, still take about 90 minutes to make the journey around the Earth once.[11]

Using the circumnavigation distance of the Earth via low Earth orbit unit on a sort-of cosmic yardstick, let us examine some intervals we tend to throw around in casual conversation. The Moon, our nearest celestial body of any significance, is a destination where man actually has ventured. It is roughly equivalent to the distance of *ten* low Earth orbits away.[12,13] This implies it would take a typical terrestrial boat more than a year and a half to reach the Moon; comparatively, it would take a modern jet airliner several weeks make the trek. A state-of-the-art space capsule on the other hand could transverse the gap to the Moon potentially in about twelve hours (although it took the Apollo astronauts as long as three days). So far, none of these scales seems unreasonable; a journey to the Moon using a purpose-built spacecraft seems to be roughly analogous to a typical intercontinental jet airliner flight back on Earth.

The Moon is a heavenly body which presents itself as almost tangibly close to us. Given a clear night, all we have to do is crane our neck skyward, and we can observe the Moon, plainly, with no special instruments required. This is mostly the exception for astronomical objects. Aside from the sun, Andromeda on clear nights, and perhaps the occasional comet, the remainder of the sky without magnification is nothing more than a featureless point of light. Yet, the perceived closeness of the Moon still tends to deceive our senses. Almost universally, depictions of the Earth–Moon system are drawn with scales which are egregiously incorrect. Take into consideration the following diagram, which places both the sizes of the Earth and the Moon, and the distances between them at their proper proportions:

●   •

Earth   *Moon*

In this depiction, to the far left, the large dot represents the Earth, and likewise the tiny dot to the far right represents the Moon. The diameter of the Earth is about 8,000 miles; the separation of the Earth–Moon system is shown to scale at approximately 239,000 miles. The Moon is the Earth's only significant natural satellite; most typical man-made low-Earth orbit satellites, such as the International Space Station, with an orbital altitude under a few hundred miles, are far too small in comparison to be represented at this scale.[11]

Beyond the Moon is where things begin to differ drastically. The distance between Earth and the planet Mars is (at minimum) about 34 million miles; though the separation is typically much larger.[14,15] In order to show the minimum Earth–Mars distance using the same scale as in the previous diagram, we would require pages thirty-two feet across! Continuing with our previous comparisons using terrestrial travel, the journey at nautical rates would be impossible; by boat, it would take more than two-hundred years – the equivalent of encircling the globe consecutively more than 1,300 times. By swift jet, it would be a slightly more tolerable ten years – though a decade's worth of airline peanuts might test one's sanity. The fastest spacecraft in development at the time of this writing – NASA's Orion Multi-Purpose Crew Vehicle, is a capsule-type spacecraft evoking the nostalgia of the Apollo program. Using Orion, it would still take more than seven months to reach Mars under the most ideal of close approach conditions.[16]

For anything farther away than Mars, it becomes entirely absurd to even consider comparing other modes of Earthly travel; rocket propulsion is the only present option if we want to reach our destination within a human lifespan. A crewed mission to Saturn (at its closest approach) aboard an Orion capsule would take nearly four years, each way. This is the equivalent of travelling approximately thirty thousand Earth orbits; to compare, the total number of orbits completed by *all* NASA space shuttle orbiters combined over their roughly thirty year lifetime was only just over twenty-one thousand.[17] Saturn, though distant, is a theoretically attainable destination using an Orion-derived spacecraft, but we are beginning to skirt the edge of what is feasible, or even possible. Using a purpose-built craft for the mission (the most logical approach) we could at best only cut this time in half via executing a Jovian gravity assist maneuver halfway through the mission (provided one could shield astronauts from the radiation environment close to Jupiter). The "ice giants" of Uranus and Neptune are significantly farther away; in the latter case, it would take about a decade to reach Neptune. If we wanted to visit the nearby star Proxima Centauri, which is around twenty-five *trillion* miles away, or an equivalent distance to a *billion* orbits around the Earth – the journey would take *over thirty-thousand years* aboard an Orion spacecraft. This is certainly a long time, despite that we are referring to the fastest human-carrying machine ever constructed – a ship capable of overflying the distance between Los Angeles and New York City in an unimaginably fast *nine minutes*.

Using our present-day state-of-the-art chemical-only propulsion technology, creating a spacecraft whose sole purpose was to reach Proxima Centauri in as little time as possible, it would still take an excruciating 20,000 years. This assumes withstanding an extremely close approach to the sun in order to attain a maximal gravitational assist. On journeys spanning many millennia, there is another important consideration – we would need to start by figuring out relative velocities between the solar system and our stellar neighbors. Just in the same manner we have to aim not where Mars is now, but where it will be in several months, so do we have to aim for where nearby stars will be in tens of thousands of years. Our journey to Proxima Centauri would be so incredibly long, that by the time we arrived, its position relative to the Earth would have changed on the order of light-years. For other relatively "nearby" targets, some of these stars which are closest to us today will not be the closest to us by the time our hypothetical spacecraft can reach them. This underscores the extreme vastness of space on even these absolute smallest of stellar scales. Think of how this would compare if our destination was not the *closest* star to the solar system, but perhaps one on the other side of the Milky Way. We would have to multiply these times and distances by *twenty-five thousand*; it would take humans half a *billion* years to make the trek, one way.

Yet, our sights are still set narrow; these distances, are mere baby steps on galactic scales; smaller still, are these steps on *intergalactic* scales. To travel to a galaxy such as our neighbor Andromeda using this mode of travel – I am saying "*like Andromeda*", as it is already moving towards us faster than we can move towards it utilizing existing chemical propulsion means – would take an eternity. Sun like stars which have not yet been formed, will have long burned out before we can reach a single one. While Orion seems fast in terms of the Earth and even possibly the Moon, in terms of not the universe, or even the galaxy, but merely our stellar backyard, it is incomprehensibly slow.

Earlier, we discussed Betelgeuse, as an extremely massive, yet relatively nearby star. It is one of the *very* few stars which are both close enough and large enough to Earth that powerful telescopes have the ability to measure their diameter directly, something known as *angular diameter*. This measurement describes the number of degrees (fractions of a degree in all cases except for the sun) of the sky which the disc of the star occupies. Other examples include Proxima Centauri, a star more than 150 times closer to Earth than Betelgeuse, and in fact *the* closest star to Earth at just over four light years away. Another is Polaris α, the principal star in the stellar system collectively known as the *North Star*, prominent in the Northern Hemisphere night sky. Even in these relatively few cases, the image resolved is far from crystal-clear; in fact even resolving more than a few blurry patches of light, a few pixels of an image, is considered a triumph. In some of these cases, sophisticated computer image processing can be applied to extract additional information in the form of a few more precious pixels; or at least additional descriptive numerical data. The point here is, that aside from these few notable exceptions – stars which are extraordinarily large or very nearby – all other stars in the *entire* visible universe are *at best*, to us, a featureless point of light. Most are nothing more than smears visible only as a consequence of their sheer quantities – billions in number – within the galaxies which contain them.

If a picture could be worth a thousand words, I know one that could be worth ten-thousand. Known to many enthusiasts, the *Hubble Ultra Deep Field*, could possibly be the most breathtaking image in astronomy to ever have been acquired. It depicts a small patch of the night sky for which its majesty is invisible to the unaided eye. One may pass it off as some tiny empty patch between the stars of *Ursa Major*. Such goes without a thought; when we look up into the darkness without a telescope, that is all we can see – vast expanses of blackness dotted by a few bright specks of light.[18]

With the exception of a few foreground stars, each tiny, minute smudge and fleck within this image is actually an entire *galaxy*, each and every one containing

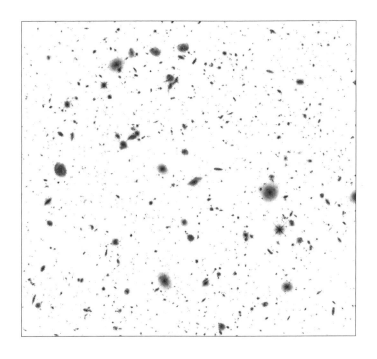

Hubble Ultra Deep Field.

Image Courtesy of NASA, ESA and
the Space Telescope Science Institute.

tens to hundreds of *billions* of stars. This photograph represents one of the farthest and deepest views into the unknown we have ever looked, an exposure taken by the Hubble Space Telescope (HST) camera with an effective exposure time of more than three weeks – all within a patch of sky smaller than a pea held at arm's length. Still, this image (shown opposite) contains upwards of ten-thousand galaxies, resolved at the limits of the capabilities of the telescope which captured them. To give a comparison of just how tiny an area this covers, it would take more than *ten million* photographs of this size to encompass the entire night sky. At Hubble's capabilities, it would also take over *eight thousand years* to capture these photographs. It is fun to imagine if we had some long-lost twin in some far-off galaxy contained within this image, who also possessed a an instrument with capabilities similar to the HST; this could just as well be their view of our home, the Milky Way being a faint smudge of light among any of the other ten-thousand or so galaxies. In fact, if Hubble had even greater imaging capabilities, it would not be unreasonable to suspect that looking further would reveal even more galaxies in the distance. The answer to this question may come sooner than one might think, when the much anticipated James Webb Space Telescope (JWST) enters service sometime in the year 2018, provided the program doesn't suffer any further delays. While the images procured by Hubble in the visible and near-visible spectrums have arguably attained the pinnacle of astronomic observations, the JWST, will undoubtedly eclipse this.

With a mirror bounding more than seven times the light collecting area as Hubble, the JWST observes primarily in the infrared spectrum. As such, it will have the capability to attain images with higher redshifts – and thus further into the past – approaching the horizon of the visible universe, and at unprecedented resolutions.[19] Under the prevailing Big Bang model of the universe, the JWST will theoretically be able to directly observe the period of time corresponding to the formation of the first galaxies, a time just after the birth of the first stars. These observations will likely definitively confirm or refute a handful of our preconceptions of the origin and structure of the universe.

Insomuch that we are beginning to get a handle on the vastness of distance, we should also address the vastness of time. Time relentlessly flows into the past. But what is *the past*? Perhaps it could be described as a history of things which used to be "now", but are currently no longer? Even reading the pages of this book, you saw them as when they were ever so slightly younger - even if it was only an imperceptible fraction of a second ago. Most objects we interact with are so near, we can completely ignore the fact that these objects exist not only at a distance in *space*, but also a distance in *time* away from us. All this can be easily changed with the simple act of tilting our heads towards the sky. When we gaze upon the Moon, we see it as it was about a second ago; the planet Venus, prominent and bright, and frequently seen in close proximity to the Moon in the night sky is on average about twenty-six million miles away – several minutes into the past. Further still, we see fainter glints, nearly indistinguishable except to the most astute eye; some of the stars you see as they were while you were growing up; a few you may see as they were when you were born. A few of these stars have died long ago, now corpses whose light still reaches across the sky like ghosts.

Nothing beyond the Milky Way is as it appears; even when it comes to our closest neighboring galaxies, reality becomes exceedingly difficult to distinguish from optical illusion. The starscape of the heavens, is in continuous flux; it is ever-changing. Though humans often perceive it as such, the night sky is not a static cross-section in time. A more appropriate analogy would be to conceptualize it as a near-infinite stack of chronologically layered snapshots, like superimposed frames of a movie reel. Most of each frame is transparent, enabling us to see stars and galaxies contained in other frames from different times. Galaxies hurtle to and fro, yet most of them appear to recede ever-faster away from us at greater distances into not just the vast reaches of space, but time as well. As a consequence of this, neither the order in which the frames are stacked is preserved, nor are the frames chronologies constant relative to each other. These heavenly vistas are also different from every possible vantage point in the universe.

Views of the most distant galaxies, actually are indeed more akin to optical illusions than representative of the universe *"live and in color"*. Yet, all one needs is a powerful enough telescope (pointed in *any* direction in the night sky) for the innumerable galaxies to become apparent. Earlier modern thinkers speculated that each might be their own "island universes"; such ideology was promulgated by the mid-eighteenth century astronomer Thomas Wright and philosopher Immanuel Kant, later to be confirmed by Edwin Hubble. Now known today is that each and every one of these galaxies contains more stars than people who have ever lived. We may see these galaxies as they were, not at the dawn of humanity – no, that was merely yesterday to these far away galaxies – but as they were at the *dawn of time*, as we so like to define it. Such an ancient past represents that which is the longest time ago from the present that we can observe, no matter how powerful or advanced of a telescope we may build.

Would it be that much of a stretch then, to question our concept of what *now*, the present, actually means? The idea of *now*, in its everyday sense, is the idea that while some event is occurring locally, there is another, distinct arbitrary event occurring simultaneously, typically at some arbitrary distance away. We imagine that *if* we were able to instantaneously transport ourselves to *that location* at the *present time*, we would be able to experience this simultaneity. In reality of course, this is not possible; we are well aware that it takes time to travel from here to there, and in that span of time, *now* no longer exists. To give an example more plainly, we are aware that our hypothesized distant cousin in China, though half the world away, is indeed performing some action *now*. Whether that action is walking down the street, or climbing a tree in the forest, or laying sound asleep, we have confidence that *the present* exists for our Chinese cousin, just the same it exists for you reading this book right now. We are well aware of the fact that to know his action, we would have to board a plane or a ship, or otherwise make some arrangement so we can discuss our cousin's day. Even in picking up the telephone and asking him to explain to us his actions, *now* is always slightly in the past, even if by that miniscule fraction of a second it takes light and electricity to transverse the telephone lines. It is true with any event, even the simple act at watching an ant crawl across the ground, we see the ant as it was the slightest fraction of a moment ago. Light needs to reflect off the ant and travel through our eyes; in the nerve from our eyes into our brain, and then the miles and miles of nerve fibers which make up our consciousness, all contribute to experiencing *now* this ever so brief moment ago. In fact, it is *only* when occupying the same spatial coordinates where *now* has any real equivalence – an impossible feat in reality of course, as prevented by the uncertainty principle. *Now* is a fictional place in time; it is our conceptual generalization of events which occur in close proximity on small scales which are important to us, as processed by our brains on a moment-to-moment basis. It is, in a casual sense, the smallest unit of our attention span.

Now applies, largely in extent, to what we see, think, and experience at the shortest durations afforded by consciousness. What we see of course, is dependent on the light which reaches our eyes. Light travels fastest through the vacuum of space, maintaining a somewhat slower velocity through matter-mediums; however, for human affairs, we can treat the speed of light through ordinary air approximately to that of a complete vacuum. Using this concept, with distance measured as how far light travels over a period of time, we can begin to gain a sense of just how inept human beings are at processing the vastness of the cosmos.

What would you consider the threshold for the moment *now*? Would it be a second, or some fraction thereof? Perhaps two seconds? Looking upwards, how far away from Earth can one expect to continue to observe simultaneity? We talk of things which occur in a second as near-instantaneous, as in *"a blink of the eye"*; but a second can be a deceptively long time when you are sitting idle counting them one by one. At a distance of one light-second, our proposed "radius of simultaneity" – that is, our concept of *now* – would extend roughly to the Moon. Any would-be lunar astronauts we could spy on, using a hypothetical, ultra-powerful terrestrial telescope, would be *at the very limit* of the present. Operating a space telescope (such as the planned JWST) in a Lagrange point orbit L_2, would place it at approximately six times the distance of the Moon; most would agree we would unambiguously be conducting observations *slightly* in the past with respect to the Earth. Communication would suffer a respectable twelve second delay in round trip time – in other words, the answer to *any* interrogation asked of this satellite would not be returned for a minimum of twelve seconds. Continuing along the same lines, an astronaut on a trip en route to Mars would not be able to answer Earth-based communications for a minimum of several minutes under the most ideal conditions, to more than half an hour at the two planets' greatest separation. As one can easily see, the concept of *now* quickly breaks down even within the inner solar system – places in the cosmos which we consider so be very near. We sometimes forget the concept of simultaneity cannot be so easily applied to worlds afar.

Moving to our closest neighboring star from the sun, Proxima Centauri, the concepts of *now* and simultaneity become even more distorted. The minimum time for information exchange at this distance is more than eight years. Even if we were able to design a space probe which could reach Proxima Centauri within the lifetime of a human being, any queries sent to the probe requiring a response would take almost a *decade* to execute. This is not a limitation of our technology; it is a limitation of the laws of physics. Simultaneity continues to diverge as we move on to the second-closest star, and so-on. A probe conducting observations near the galactic core would take fifty millennia to respond to communications from Earth. For Andromeda, our closest neighboring major galaxy, such a round-trip signal transmission would take approximately *five million years.*

As clever humans, we have already identified this problem even in our short, infantile steps around our own celestial backyard. Modern space probes operate with a certain degree of autonomy, making their own adjustments in camera positioning, and, in cases, even the overall trajectory of the spacecraft without direct human intervention. The *Mars Science Laboratory*, or as it is more commonly referred to as the *Curiosity* rover, is a recent example which needed to exhibit this autonomy in navigating the red planet.[20] It would be impossible to conduct a precision rocket-controlled approach and landing on the surface if such a landing were to be controlled terrestrially. It would also be tedious and completely infeasible for a human to "drive" the rover from Earth, with the two-way communications turnaround time routinely exceeding twenty or thirty minutes. Imagine the difficulties in driving a car via remote control from a camera in the driver's seat, where the video is delayed by twenty to thirty minutes. Imagine how careful and how slow and how purposeful an operator would need to be; especially given the fact that repairs of any kind are an impossibility. Even if we sent astronauts to Mars to drive their own rovers and conduct their own experiments, it would be exceedingly difficult to carry on conversations with them in real time; the Martian crews would necessarily need to exercise autonomy from mission controllers home on Earth.

And yet, such troubles are described for a meager average distance of a hundred million miles or so. Looking towards the nearest of stars, this gap quickly turns into not billions but tens of *trillions* of miles, making the communications delay measurable in terms of not minutes, or days, or even months – but decades. The furthest stars in the galaxy require millennia; for our nearby galactic neighbor Andromeda, the communications delay burgeons to several million years; for the most distant observed galaxies, we find this gap widens to more time that prevailing theories have given the universe to exist.

As a testament to the limitations imposed by this vast space, for any significantly farther distance away from us than Andromeda, even the most powerful telescopes have difficulty separating individual stars in galaxies. For any given one of the hundred-something billion galaxies our current technology has the ability to resolve – save two, Andromeda, our own, and a few dwarf galaxies in between – we must be content with the dim foggy wispy streaks as the only indication of the hundreds of billions or more of stars each contains, with *one* notable exception. At the end of a star's life, if it is massive enough, it will flash in a moment of magnificent brilliance as it burns out. Deemed *supernovae* – intense outbursts of energy as a star remodels itself – can for but a brief moment outshine their entire host galaxy. Back on Earth, if we happen to be looking in the right direction, at the right time, we can observe this signature, of a single, solitary star, from literally almost halfway across the universe. For this brief moment, we know of this star, and a little bit of its life which it has lived its life long ago, in a place far away. How interesting to think, that for millions, or even billions of years, it was from the vantage point of other planets, their sun, shining light upon them day after day for eons.

As of the time of this writing, the most distant supernova we have ever observed occurred in a galaxy almost ten billion light-years away.[21] We know that the sun is at least a second-generation star, meaning that long ago, it was formed from the debris of a dead star which met a similar violent and cataclysmic end.[22] The evidence continues to persist all around us, in the ratios of heavier elements which are only produced in a type of massive star which burns brightly. Heavier elements still, such as lead and platinum and gold and uranium, elements critical to our modern lifestyle, can only be formed in the significant, useful quantities we possess during the catastrophic demise of such a star which ends in a supernova.[23]

Take this most distant corpse of a star, which violently exploded almost ten billion years ago. Let us use our imagination, to perform a thought experiment. If we today embarked on a journey to reach this star – travelling just under the speed of light – we would likely witness something spectacular unfold. Due to relativistic effects, time for us would slow as compared to our destination; we would perceive the voyage taking mere months, yet witness the accelerated transpiring of events at a distance. As such is the cycle of stellar life and death, from the ashes of this supernova, a new generation of stars would coalesce; perhaps one would be near one solar mass. Necessarily, we continually correct our course to following this newborn sun as it wanders in step with its parent galaxy, sailing through the universe. Approximately halfway there, we turn to look back; we depressingly notice that our own sun has consumed the Earth, and shed much of its mass to become enshrouded by a tragically beautiful diffuse nebula.[24] Turning ahead, we see a new solar system has formed, with planets resembling what was once familiar to our own. However, the most somber event of all happened just prior to our arrival at the system. As we are slowing down to explore new worlds, the star we have witnessed coalesce from the original supernova, has died and consumed many of its planets, itself becoming a planetary nebula. All that is left of the star and its terrestrial worlds are gas and dust. Our only solace is found in that perhaps one day, these remains will go on to be incorporated into a new generation of stars, as the cycle continues.

Looking deeply into the cosmos certainly has peculiar implications. Consider any distant supernova some several billion light-years away; for such an event we could very well be witnessing the chronicles of an analogue to our own solar system, unfolding before us. Here, now, we would observe these events which occurred when our own solar system appeared just the same. In these places, at this moment, in the wake of a massive supernova, there are undoubtedly *"new"* stars being formed – *new* in the sense that they are yet unseen to us, but *now* which are as old – or even much older – than the sun. What we observe as unborn stars, by this time likely have acquired their own planets in orbit around them, and brightly shine their light onto these new, unseen worlds. All of the Earth and the solar system's history, from its coalescing from the gas and dust of an ancient supernova, through everything mankind has done and accomplished; all of this time has already gone by, and perhaps then a bit more, on some planet in this faraway place, to which we cannot yet see that it even *exists* yet due to the vast separation of space and time. This vastness, in words alone, cannot be understated, nor understood.

1   Torres, G.; *et. al.* "Accurate masses and radii of normal stars: Modern results and applications". (2010). Astronomy & Astrophysics Review. (Vol. 18, 67 – 126).

2   Waller, W. H. "The Milky Way: An Insider's Guide". (2013). Princeton University Press. (264).

3   Polakis, T. "Orion the Hunter". (February 2004). Astronomy Magazine. (74 – 79).

4   Lang, K. R. "Essential Astrophysics". (2013). Springer Science & Business Media. (307).

5   Kervella, P.; *et. al.* "The close circumstellar environment of Betelgeuse". (2016). Astronomy & Astrophysics. (Vol. 585).

6   Dolan, M. M.; *et. al.* "Evolutionary Tracks for Betelgeuse". (2016). The Astrophysical Journal. (Vol. 819, No. 1).

7   "Dawn at Ceres" Press Kit. (March 2015). National Aeronautics and Space Administration (NASA), Jet Propulsion Laboratory.

8   Barnett, R. A.; *et. al.* "Analytic Trigonometry with Applications". (2011). John Wiley & Sons. (7).

9   "Virgin Atlantic GlobalFlyer". (2014). Scaled Composites, LLC.

10   Moll, N. "Globally Fastest". (1993). Flying Magazine (Vol. 120, No. 1).

11   Reibeek, H. "Catalog of Earth Satellite Orbits". (2009). National Aeronautics and Space Administration (NASA).

12   Anchordoqui, L. A. "Lectures on Astronomy, Astrophysics, and Cosmology". (2016). Lehman College, Department of Physics and Astronomy.

13   Katz, A.; Franco, M. "Targeting the Moon". (June 2011) IEEE Microwave Magazine.

14 "The International Exploration of Mars". (1993). 4th Cosmic Study of the International Academy of Astronautics (IAA), Committee on International Space Plans and Policies, Subcommittee on the International Exploration of Mars.

15 Gingerich, O.; Henry, R. C. "Planetary Pretzels: Three diagrams showing the inner workings of the orbits of Mercury, Venus, and Mars.". (November 2005). Sky & Telescope.

16 "Orion Multi-Purpose Crew Vehicle". Fact Sheet. (2014). Lockheed Martin Corporation.

17 "Space Shuttle Era Facts". (2011). National Aeronautics and Space Administration (NASA), Kennedy Space Center.

18 "The Hubble Ultra Deep Field". (2008). National Aeronautics and Space Administration (NASA), Goddard Space Flight Center, and the Space Telescope Science Institute.

19 "James Webb Space Telescope". (2015). National Aeronautics and Space Administration (NASA).

20 "Mars Science Laboratory / Curiosity". (2015). National Aeronautics and Space Administration (NASA).

21 Cooke, J.; et. al. "Super-luminous supernovae at redshifts of 2.05 and 3.90". (2012). Nature. (Vol. 491, No. 228).

22 Gounelle, M.; Meynet, G. "Solar system genealogy revealed by extinct short-lived radionuclides in meteorites". (2012). Astronomy & Astrophysics. (Vol. 545).

23 Qian, Y. "Some nuclear physics aspects of core-collapse supernovae". (1998). Second Oak Ridge Symposium on Atomic and Nuclear Astrophysics.

24   Schröder, K.-P.; Smith, R. C. "Distant future of the Sun and Earth revisited". (2008). Monthly Notices of the Royal Astronomical Society. (Vol. 386, No. 1, 155 – 163).

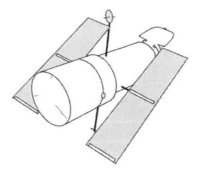

V

Hubble's Law and the Hubble Flow

"At the last dim horizon, we search among ghostly errors of observations for landmarks that are scarcely more substantial. The search will continue. The urge is older than history. It is not satisfied, and it will not be oppressed." [1]

– Edwin P. Hubble

FOR MOST OF US IN THIS DAY AND AGE, the notion that our universe is expanding on a cosmic scale seems to be relatively common knowledge. However, the exact implications of this theory admittedly remain more obscure. Astronomers and cosmologists – those scientists who have dedicated their professional careers to studying the heavens – use a certain value deemed

Hubble's constant to more precisely describe the mysterious expansion of empty space.[2] Presently, such a measurement is extraordinarily difficult to ascertain, attributable to its value being close to the quantifiable tolerances of even the most precise instruments mankind has yet conceived. Obtaining an accurate value has remained just at the horizon of possibility since we have, many decades ago, realized the medium of space-time is perpetually expanding.

Hubble's constant is appropriately named after the now famous twentieth-century astronomer Edwin P. Hubble. In the late 1920's, while observing what appeared to be faint, distant galaxies, Hubble noticed a curious optical anomaly. He noticed that the emission spectra of light originating from these allegedly distant galaxies seemed to be "shifted" towards lower frequencies – that is, shifted towards the *red* end of the spectrum when viewed in visible light. This effect, appropriately known as *redshift*, is the stretching of an electromagnetic wave (in this case, visible light) along its direction of travel, and occurs whenever there is a difference in *relative* velocity between a source of light and an observer – a consequence now known to be due to relativistic effects on space-time. For objects moving away from each other, when viewed in the visible spectrum, this effect causes the frequency of light to decrease. Theoretically, if the disparity in relative velocity were to be pronounced enough, a shifting of the perceived color towards the red end of the spectrum would occur. The object being viewed would literally appear to be redder in color; thus the phenomenon is aptly named a *redshift*.

A formal description of this effect was first set forth by the nineteenth century physicist Christian Doppler, and as such, is widely known as the *Doppler effect*.[3] The effect's occurrence is not limited to electromagnetic waves, but waves in any medium in general. Sound demonstrates an analogous, common everyday occurrence of this phenomenon. Such may be conceptualized in how, from the perspective of an observer, a pitch change is perceived from an approaching or receding source.

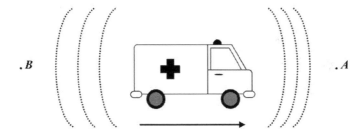

The sound of an ambulance siren is distinctive enough that it finds ubiquitous use as an example thereof; additionally, the siren is well suited for demonstration, as pitch changes in its continuous drone are obvious to a listener. Here, as the ambulance approaches an observer located at point A., the sound waves appear to "bunch up". As a result of their shorter wavelength, the perceived pitch is heard as *higher*, as shorter wavelengths result in a higher frequency sound. Conversely, as the ambulance recedes away from an observer positioned at point B., sound waves appear to become more "spread out", and the pitch is perceived as *lower*. Though our eyes are not nearly sensitive enough to detect this, electromagnetic waves in the form of light would also demonstrate the same transformation. If our eyes were somehow sensitive enough, we would strangely enough notice a shift in the *color* of the ambulance. From the perspective of point A., the ambulance would look *bluer*; and conversely from the perspective of point B., it would appear *redder*.

While in the above example it would be difficult to measure on such small objects and small relative velocities, the electromagnetic Doppler effect becomes pronounced enough on astronomical scales to yield useful information. When astronomers employ it to spectrographically analyze the sun, the Doppler effect yields a measurable difference between the advancing and receding "edge" of the solar disc as it rotates from the perspective of the Earth.[4] The light when viewed through a prism contains narrow, sharply dark bands in the otherwise smooth continuous spectrum of color. These dark bands, otherwise known as *absorption lines*, are due to the quantum interactions between photons and atomic orbitals in the matter from which the observed light passes through. In redshifted light, these absorption lines are also shifted towards the lower frequency, or "red" end of the spectrum. Likewise blueshifted light has these absorption lines more towards the higher frequency, or "blue" end of the spectrum. These dark lines serve as a sort of fingerprint of the material, indicating its composition – that is whether it is made of hydrogen or helium or sodium, *etc.* Emission spectra provide similar insight. In the context of the sun, we are able to observe this minute shift in the absorption lines from side to side across its equator, not only telling us the general composition, but also the precise rate at which it rotates. Modern astronomers use redshift (and the reciprocal effect of *blueshift*) in order to determine a range of properties related to distance and relative velocity of stars, nebulas, galaxies and other heavenly bodies.

Applying knowledge of the Doppler effect, Hubble made an incredible observation. No matter what direction he pointed his telescope, he observed that virtually *all* distant galaxies were receding away from us. Even more astonishing, was the fact the redshift effect was more pronounced, the fainter and farther away the galaxy appeared.

Hubble was able to discern that distant galaxies were progressively becoming more distant, with galaxies further away becoming more distant at a *faster* rate. He was able to correlate the distance to far away objects in the cosmos was proportionally related to how fast it was receding away from us – he found that the most distant objects receding away from us the fastest. The only

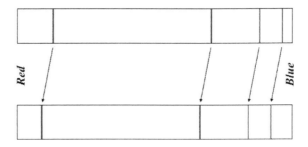

A typical schematic for the redshift in the spectrum of hydrogen. The object (*e.g.*, a star) possessing the upper spectral signature is moving toward an observer with respect to the lower signature.[5]

logical explanation which could be offered was that the universe, as a *whole*, was expanding. For Edwin Hubble, as an individual who had dedicated his professional life to studying the cosmos, I can only begin to imagine how exciting a moment it must have been when he came to such a realization.

Hubble's constant can be a source of incredible amazement. Mathematicians have empirical expressions involving fundamental constants, such as the ratio of π, or the natural logarithmic base, e; physicists likewise have their own analogues. The speed of light, c, is fairly high on the list, among others such as Planck's constant, h, and the universal gravitational constant, G.[6] Hubble's constant – which we denote as H_0 – is a relative newcomer in this regard, as only recently has science been able to ascertain its value, and only more recently measurable with any precise certainty. It is rapidly becoming apparent however, of the intrinsic nature of Hubble's constant in explaining some of the most fundamental qualities of our world.

Owing to its difficulty in measuring, the range of values attributed to Hubble's constant have gradually narrowed over the years as increasingly more advanced technology has become available. The primary barrier in obtaining a true measurement of H_0 is obtaining precise distances to far away, significantly redshifted targets (*e.g.*, distant galaxies). Scientists have to resort to clever tricks, such as the predictable properties of variable stars (*e.g.*, *Cepheids*) and type I_a supernovae. This is accepted under the likely pretense that the laws of physics are universal – continuing to follow that same revelation that the same laws of physics on Earth also apply to distant galaxies. To the extent we have seen of the visible universe, this appears to be the case.

Today, the most precise measurements have come from space missions which began with NASA's Cosmic Background Explorer (COBE) mission in 1989, alongside observations conducted using the Hubble Space Telescope (HST) since it was launched in the early 1990s. This was followed by the more recent NASA Wilkinson Microwave Anisotropy Probe (WMAP) survey, finally culminating with the joint European Space Agency/NASA Planck mission in 2013.[7,10] As the names of these missions suggest, the primary manner in which data was obtained consisted of building a survey of the cosmic background radiation, and examining that data for clues to the origin and nature of our universe. The best currently measured value for Hubble's constant, H_0, is 67.8 ± 0.9 km s^{-1} Mpc^{-1}, as reported using data up to and including the most recent studies.[10,11] Depending on how the data is analyzed and interpreted, there remains some minor disagreement with the generally accepted value, however it is better by an order of magnitude over estimates prior to integration of the newly available Planck mission data.

Briefly describing the general approach to The Planck mission can give some indication of the difficulty ascertaining Hubble's constant – a primary objective.

Hubble's Constant (H_0)
$km \ s^{-1} \ Mpc^{-1}$

Mission	Year	Value of H_0		
Planck[10,11]	2015	67.8	±	0.9
Planck[10]	2013	67.3	±	1.2
WMAP[8,9,10]	2009	69.32	±	0.8
HST[12]	1999	72.0	±	8.
Prior accepted value[13]		75.	±	25.

Measurement of the rate of expansion of the universe demonstrates a progressive convergence culminating in the most recent data collected inclusive of the Planck mission.

The atmosphere of Earth is largely opaque to microwave radiation, necessitating the use of a space-based instrument to obtain precise measurements. Even in the cold, quiet isolation of space, electromagnetic interference remains a concern; the Planck spacecraft was to conduct its observations in a heliocentric (that is, *sun-centered*) orbit about the Earth-Sun Lagrange point L_2. This places the probe in an arrangement where Earth always remains situated between the spacecraft and the sun. A distance of approximately 1½ million kilometers ensures that most interference of terrestrial origin is mitigated; likewise, a great deal of interference otherwise caused by the sun is attenuated in the shadow of the Earth.

Schematic depicting the *Lagrange* point (L_2) orbit location of the ESA/NASA Planck Mission (not drawn to scale).

Onboard, a reservoir of cryogenically cooled liquid helium was used to reduce the temperature of its instruments down to 0.1°K – that is, just a fraction of a degree above absolute zero – in order to remain sensitive enough to detect the faint background radiation travelling billions of light years across the cosmos. When operational, the background radiation detectors aboard Planck were popularly reported to be *"the coldest known object in space"*, natural or artificial.[14]

This necessary arrangement underscores the prerequisite great feats which must be realized in order to accurately and precisely determine H_0. Additionally apparent is historically why, in the absence is sufficiently advanced technology, the empirical nature of Hubble's discoveries were not obvious. It is only after preceding frameworks matured; after we have launched probes deep into space, far away from Earth and at great expense of time, effort, and economy, that Hubble's constant resolves to a number which is accurate enough to be meaningful and useful.

The actual equation for Hubble's Law is a deceptively simple expression, merely a certain velocity over a distance; accordingly, we may succinctly write Hubble's Law as:

$$H_0 = \frac{v}{d}$$

Particularly, v is the *recessional* or "radial" velocity of a distant object from an observer; d is simply the distance to that object. Hubble's constant is represented as H_0, deriving its small subscript zero as an indication we are here referring to the present-day value of the constant with respect to the history of the universe. This is done for clarity with regard to other theories which allow the value to change over time – for which of course the descriptor "constant" would become a misnomer.[15]

As it turns out, Hubble's constant divulges a treasure trove of information, describing to us multiple crucial aspects of the cosmos. First and most obviously it plainly states the rate of expansion per unit volume of space. Astronomers, dealing with the vast distances separating anything interesting, use a measurement of distance known as the *parsec*, or "*parallax of one arc second*". The description of distance in this manner dates back more than a hundred years ago, when astronomers first required a means to describe large, interstellar distances. It is perhaps best to conceptualize the distance in terms of the more common light year; in this regard a single parsec is equal to approximately 3.26 light years. The star *Proxima Centauri*, our nearest stellar neighbor after the sun, can be said to be an estimated 4.25 light years away; alternatively, we can say the distance to this nearby star is 1.3 parsecs.[16]

Consequently, the *megaparsec* (Mpc) is a vast unit of length equivalent to one million parsecs (again, conceptually, about 3.26 million light years). Thus the value of H_0 at approximately 67.8 km s^{-1} Mpc^{-1} implies that every linear megaparsec of space – that is each distance of 3.26 million light years expands by about 67.8 kilometers every *second*. While 67.8 kilometers (roughly forty-two miles) may seem like a fairly significant distance just to "manifest" itself into existence, over the scales of millions of light years, it ends up being rather insignificant to any would-be inhabitants (*e.g.*, us) within that space. One light year is a distance spanning more than *nine trillion* kilometers. Comparatively, 3.26 million light years represents a distance more than thirty-two times the estimated diameter of our entire galaxy.[17] Therefore, contrasted to 67.8 kilometers, we find the expansion only equivalent to about seven *ten-billionths* of one percent, an incredibly small value in comparison. The effects of gravity and the other fundamental forces are usually sufficient to make the effects of space-time expansion undetectable on anything shorter than *intergalactic* scales. Objects separated by vast distances of space and not gravitationally bound to any appreciable degree, will observe the same corresponding increase in velocity (of ~67.8 km s^{-1}) away from each other over one second, increasing as the objects move further apart and the distance between them grows larger and larger.

An unfortunate logical fallacy that has developed as a result of the incorrect promulgation of this theory, wrongly explains that as space-time expands, the contents *within* space also expand. Simply put, this is false. Stars and galaxies do *not* grow and become larger as a consequence of the relentless expansion of space – they maintain their volumes, mass, and other physical properties, and more or less continue along their local trajectories as if space was static. Only at distances where the gravitational interaction is too weak between bodies to "close" the space at it expands is the phenomenon discernable.

A commonly used analogy is that of a baking loaf of raisin bread. We can think of the "dough" to represent the medium of space-time, whereas we can think of the "raisins" to represent stars, galaxies, planets, and other bodies which it contains.

Like a rising loaf of raisin bread, only the medium of
space-time expands; contents retain their original geometries.

As the loaf bakes, it raises, expands, and the raisins – like the galaxies they represent – recede from one another. However, the raisins themselves, remain essentially unchanged; only the distance between them has increased, because the medium, the dough, has expanded. But unlike the rising loaf, the expansion of our universe never ceases. To the extent of what we today know, more and more space is continually manifested, the cosmos becoming ever more voluminous. From the perspective of any fixed point, one may perceive this as a sort of "flow", not unlike an imaginary river which continually accelerates ever faster. In the schematic below, a typical observer is located at point *P.*, here looking leftward. At increasing distances, the relative velocity between the observer and a distant object exponentially increases. Very distant objects, such as far away galaxies, are moving away briskly. As the distance becomes sufficiently large enough, the corresponding rate of recession approaches the speed of light. The schematic below illustrates the widening topography of space-time from the perspective of an observer at a hypothetical point *P.*

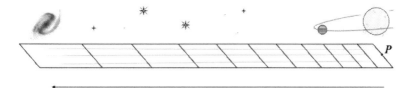

The expansion of the universe, or the "*Hubble flow*", extends radially in all dimensions from any observer at a "fixed" point.

While it is true, universally, that no object may travel faster than the speed of light, there is no law of physics which prevents apparent *relative* velocities between two objects separated by vast distances of space to transcend this limit, as a result of the perpetual expansion of space-time. Along a geometrically straight line in any direction from an observer at a fixed point, at a sufficient distance, expanding space will carry away distant objects faster than the light can transverse the widening gap back to the observer. Beyond a certain distance, the relentless expansion of space will forever separate us from the physical universe beyond such. Appropriately named, this describes one type of an *event horizon* – in this case, located at the outer edge of the observable universe.

Galaxies observed approaching such great distances away appear markedly redshifted. Astronomers use a parameter, denoted z, to represent this relative Doppler redshift, which can be concisely quantified via the following expression:

$$z = \frac{\Delta\lambda}{\lambda}$$

The wavelength of observed light is represented by λ; therefore, the z parameter simply describes the relative difference between the observed wavelength of light, as compared to the expected wavelength. By comparing recorded spectra (and the location of absorption or emission bands) of distant heavenly objects in motion to known terrestrial spectra, the relative velocities of all cataloged objects can be ascertained. For example, an object possessing z = 0 would indicate no Doppler shift, and that the relative velocity to such an object is nearly zero. Conversely, z > 0, or positive values, would represent an object moving away from an observer, and z < 0, negative values, would indicated an object moving towards an observer.[18]

Just as Hubble noted, the vast majority of objects, at large distances, are frequently moving away. Furthermore, at a sufficiently large distance, *all* objects are *always* moving away. Astronomers are continually searching for objects exhibiting increasingly high values for z, as these represent objects near the edge, close to the event horizon of the visible universe. These observations serve to validate (or potentially refute) our knowledge of the universe as we presently understand it.

In early 2016, it was announced that a distant galaxy GN-z11, in the vicinity of the "Big Dipper", was observed using the Hubble Space Telescope which exhibited a redshift quantified at $z = 11.1$.[19] Recall that as we look at objects which are very far away – on the order of tens of billions of light years – we are also looking very far into the past. Current estimates under the generally accepted ΛCDM[†] model of the universe constrain us to a finite lifetime, some estimated 13¾ to 14½ billion years or so. Therefore, observations such as this are significant in that not only do they occur near the edge of the visible universe, but also just as strangely near the beginning of time itself.[20,21] How peculiar a notion it is, that we may gaze nearly to the beginning of creation.

With respect to the understanding of the universe, the fact that we are witnessing an entirely formed galaxy – apparently filled with stars, and a luminous central black

---

[†] The "*lambda cold dark matter*" (ΛCDM) model represents a generally accepted prevailing theory describing the origin and structure of space-time and the contents of the known universe.

hole – is particularly vexing. The lifetimes of stars are measured on the order of – at minimum – tens of millions of years, and more typically in the case of stars like our own sun, billions of years.[22,23] Given its chronological placement close to the origin of the universe, the prevailing understanding of the cosmos does not afford GN-Z11 very much time to organize its galactic structure, coalesce stars, or assumingly accrete its luminous supermassive central black hole.

Considering the aforementioned observation of GN-Z11, determined to be so far in the past, we are brought to another important revelation of Hubble's constant. If the obvious quantity if a rate of expansion per distance, a second, slightly more obscure feature of Hubble's constant comes from its expression of a value in units of distance *per* time *per* distance – something which we can simplify units to that of inverse time, s^{-1}. Converting from units of km s^{-1} Mpc^{-1}, this yields a value of approximately 2.18×10^{-18} s^{-1}. If we consider the term H_0^{-1} (the reciprocal of Hubble's constant), we end up with a term expressing time in seconds. With some further conversions, we can more conveniently express in years:

$$\frac{1}{H_0} \approx 4.59 \times 10^{17}\text{s} \times \frac{1\text{d}}{86{,}400\text{s}} \times \frac{1\text{yr}}{365\text{d}} \approx 14\tfrac{1}{2}\text{ Gyr}$$

The reciprocal of Hubble's constant equates to about 14½ billion years, a result which roughly coincides with current estimates of the age of the universe based on leading cosmological models. This is in fact not a coincidence, but rather resembles the actual calculation used to determine the age of the universe. Most widely promulgated estimates on the age of the universe derive as a result from calculations which multiply H_0 by an arbitrary constant with a value close to one, to more closely coincide with experimental astronomical observations. The removal of this contrived "fudge-factor" achieves better agreement with the preceding expression. Such may be a clue that a disparity exists between our conceptions and reality, which is worth entertaining.

Data obtained from the recent Planck mission and its predecessors mounts strong evidence in support for the family of cosmological models under the umbrella term of "Big Bang" theories (including the prevailing ΛCDM model) which imply in one form or another that the universe arose from something resembling a point, subsequently expanding to the universe we observe today. The logic in this model is such that, given the universe is expanding, conceptually as we run time backwards, the universe contracts, and as such may only contract so far before compacting itself into a single point. This is a clever idea; however, it has certain implications which are not very elegantly explained. Accepting this model suddenly places constraint on everything there is and ever was, and does so on a timescale for which our planet Earth as existed for a significant portion of such (this is, roughly 4½ billion years of the approximate 14½ billion years since the beginning of space, time, and the universe). In the vastness of the cosmos, to state our planet has existed for such a significant part of everything there ever was, seems to be a dubious proposition.

Throughout physics and astronomy, Hubble's constant appears in numerous models and equations describing the universe on the largest of scales. Renowned physicist Alexander Friedmann during the 1920's put forth a number of interrelated equations (many of which can be considered different forms of the same equation) describing the expansion of the universe on vast scales.[24,25] Assuming a flat universe (where the curvature, k, is modeled as zero), and likewise a density parameter equal to one (Ω_0 = 1), we can make substitutions to Friedmann's work, with inclusion of Hubble's constant, to arrive at the following:

$$\rho_c = \frac{3H_0^2}{8\pi G}$$

Here, we are presented with ρ_c, the *critical density parameter*, which is defined as an expression of fundamental constants. H_0 is of course Hubble's constant, here appearing squared; the term G is again the *gravitational constant*, and is equal to approximately 6.67384×10^{-11} m^3 kg^{-1} s^{-2}.[26] Describing ρ_c as *critical* pays homage to how the universe has been regarded up to this point – originating from a singularity, destined to expand forever, dramatically collapse upon itself, or neither at precisely a certain critical density. This parameter ρ_c is important because it describes the distribution of matter on scales large enough that it becomes approximately homogenous, thus allowing us to describe the manner by which the universe operates in a general way. This is much in the way other branches of science can generalize the properties of gasses and fluids to make generalized calculations to a high degree of accuracy and precision (*e.g.*, applications of the ideal gas law, computational fluid dynamics, *etc.*). As a matter of fact, the observed homogeneity is so consistent on very large scales, its behaviors can actually be remarkably approximated and modeled as if it *actually were* a gas or a fluid.[27]

The critical density parameter ρ_c as we have previously mathematically defined describes a universe which is topologically flat; this is supported from current observations, whereby the density of the universe appears to be close to or equal to ρ_c on the grandest of scales.[10] If we continue with the assumption that the density of the universe approximates or is equal to the critical density ρ_c, knowing the volume of the universe would enable us to relatively straightforwardly calculate the mass contained within.

Given that we have shown that the reciprocal of Hubble's constant can be expressed as a period of time, which has been dubbed *Hubble time*, we can likewise ascribe a parameter to represent a distance transversed by light within such an interval. Similarly, *Hubble length*, represented by the term r_{hs}, is the distance traveled by light in *Hubble time*. Therefore, we would expect the radius of the visible universe to be:

$$r_{hs} = \frac{c}{H_0}$$

The expected duration of time for how long the universe has been expanding based prevailing models of cosmology might also be referred to as the *Hubble age*. As such, at scales representing the entire universe, time may be expressed in units of Hubble time, and distance in units of Hubble length.[28] Using the formula for the volume of a sphere consisting of radius r_{hs}, and making use of the above substitution, we arrive at:

$$V_{hs} = \frac{4}{3}\pi \left(\frac{c}{H_0}\right)^3$$

In this above expression, r_{hs} is, again, the Hubble radius, which can be said to describe a *Hubble sphere* centered on an observer; and V_{hs} is the volume, effectively the volume of the visible universe from the perspective of said central observer. Accordingly, with the ability to calculate the Hubble sphere volume, we can multiply by the critical density ρ_c to arrive at the mass contained within our visible universe:

$$\rho_c \times V_{hs} = \frac{3H_0^2}{8\pi G} \times \frac{4}{3}\pi \left(\frac{c}{H_0}\right)^3 = M$$

This gives us M, expressing the mass contained within our visible universe in kilograms. Performing the calculations, we arrive at a value of approximately 9.1877×10^{52} kg. We are confronted with the supposition that the mass of the visible universe – no matter where the observer happens to be centered within the universe as a whole – is static, a fixed quantity dependent upon other universal constants of physics.

So, rather than the *absolute* age of the universe being this thereabout 14½ billion years, it would therefore be more appropriate to refer to the universe as a continuum of "epochs" lasting for this time frame. A hypothetical particle outside the influence any fundamental force would be expected to leave the observable universe within this time, from such a the fixed vantage point. At the same time, the continual expansion of space via the Hubble flow maintains a mass fixed at about 9.1877×10^{52} kilograms in accordance with maintaining ρ_c.

1   Mayall, N. U. "Edwin Powell Hubble". (1970). National Academy of Sciences. (175 – 214).

2   Freedman, W. "The Hubble Constant and the Expanding Universe". (2004). Sigma Xi, The Scientific Research Society. American Scientist. (Vol. 91).

3   Wolfschmidt, G. "Christian Doppler (1803 – 1853) and the impact of the Doppler effect in astronomy". (2003). Hamburg University, Institute for the History of Science.

4   Gill, R. M. "The Rotational Period of the Sun Using the Doppler Shift of the Hydrogen alpha Spectral Line". (2012). Society for Astronomical Sciences, 31st Annual Symposium on Telescope Science. (225 – 227).

5   Burkhardt, C. E.; et. al. "Foundations of Quantum Physics". (2008). Springer Science & Business Media. (6).

6   "Fundamental Physical Constants – Extensive Listing". (2002). United States Department of Commerce, National Institute of Standards and Technology (NIST).

7   Boggess, N. W.; et. al. "The COBE Mission: Its design and performance two years after launch". (1992). The Astrophysical Journal. (Vol. 392, 420 – 429).

8   Bennett, C. L.; et. al. "The Microwave Anisotropy Probe (MAP) Mission". (2003). Astrophysical Journal. (Vol. 583).

9   Bennett, C. L.; et. al. "Nine-Year Wilkinson Microwave Anisotropy Probe (WMAP) Observations: Final Map and Results". (2012). The Astrophysical Journal Supplement Series. (Vol. 208, No. 2).

10 Planck Collaboration. "Planck 2015 Results. XIII. Cosmological parameters.". (2016). Astronomy & Astrophysics, Manuscript.

11 Lawrence, C. R. "Planck 2015 Results". (March 2015). National Aeronautics and Space Administration (NASA), Astrophysics Subcommittee.

12 Freedman, W. L. "Final Results from the Hubble Space Telescope Key Project to Measure the Hubble Constant". (2000). The Astrophysical Journal. (Vol. 553, 47 – 72).

13 Savage, D.; et. al. "Hubble Completes Eight-Year Effort to Measure Expanding Universe". (1999). National Aeronautics and Space Administration (NASA), Goddard Space Flight Center, and the Space Telescope Science Institute. *Press Release.*

14 "Coldest Known Object in Space is Very Unnatural". (July 2009). Space.com. Retrieved December 2016.

15 Deza, M. M.; Deza E. "Distances in Cosmology and Theory of Relativity". Encyclopedia of Distances. (2016). Springer. (578).

16 Ince, M. "Parsec". Dictionary of Astronomy. (1997). Taylor & Francis.

17 "Size and Scale of the Universe". (n.d.). Wide-field Infrared Survey Explorer (WISE). University of California Berkeley.

18 Illingworth, V. "Redshift". Macmillan Dictionary of Astronomy. (1985). Springer. (314).

19 Oesch, P. A.; et. al. "A remarkably luminous galaxy at $z = 11.1$ measured with Hubble Space Telescope grism spectroscopy". (2016). Astrophysics Journal. (Vol. 819, No. 2).

20 Ostriker, J. S.; et. al. "Cosmic Concordance". (1995). Princeton University, Department of Astrophysical Sciences.

21 Christiansen, J. L.; Siver, A. "Computing Accurate Age and Distance Factors in Cosmology". (2012). California Polytechnic State University.

22 Bromm, V.; Larson, R. B. "The First Stars". (2004). The Annual Review of Astronomy and Astrophysics. (Vol. 42, 79 – 118).

23 Larson, R. B.; Bromm, V. "The First Stars in the Universe". (2001, 2004). Scientific American.

24 Friedmann, A. "On the Curvature of Space". (1922). European Journal of Physics (formerly, *Zeitschrift für Physik*). Translated by Doyle, B.

25 Uzan, J., Lehoucq, R. "A Dynamical Study of the Friedmann Equations". (2001). European Journal of Physics. (Vol. 22, No. 4).

26 Mohr, P. J.; *et. al.* "CODATA recommended values of the fundamental physical constants". (2012). Reviews of Modern Physics. (Vol. 84).

27 Iqbal, N.; *et. al.* "Thermodynamical Model of the Universe". (2012). Electronic Journal of Theoretical Physics. (Vol. 9, No. 26, 269 – 276).

28 Longair, M. "Galaxy Formation". (2007). Springer Science & Business Media. (342).

VI

DARK MATTER,
DARK ENERGY,
AND BIG BANGS

"In terms of the most astonishing fact about which we know nothing, there is dark matter, and dark energy. We don't know what either of them is. Everything we know and love about the universe and all the laws of physics as they apply, apply to four percent of the universe. That's stunning." [1]

– an interview with Neil deGrasse Tyson

DARK MATTER, DARK ENERGY, AND BIG BANGS. These are three terms, which on the grandest of scales demonstrate that – despite our great efforts and monumental achievements in bettering our understanding of our world – we still know very, very little. Science is literally pervaded with discoveries over the ages where explanations are given that are not necessarily completely correct; however many of these partial solutions often serve as invaluable stepping-stones to advancing our understanding of our world as a whole. Consider the study of alchemy, a primitive science historically encumbered by innumerable half-truths and mistruths. Despite its inherent fallacies, it ultimately was responsible for giving rise to arguably one of the most ubiquitous and important scientific disciplines of the modern era – *chemistry*.[2] It would stand to reason, that a field in its infancy such as cosmology may likely contain explanations of the universe which are certainly not infallible doctrine, but rather more likely instead close approximations and analogues. Such shall be continually refined as our understanding evolves and matures over time. After all, the nature of science poises itself at best to only ever hope to asymptotically approach the ultimate truth.

In contemporary times, leading cosmological models – explaining the universe, everything in it, and the origin thereof – are well-popularized under the "Big Bang" scenario. This itself is an umbrella term for a collection of related theories, in which we are presented with the universe beginning as an infinitesimally small point from which, over an infinitesimally length of time, all matter, energy, space and time spring forth into existence. Previously introduced, the ΛCDM model is the prevailing theoretical framework which, to the extent of our present understanding, best explains the behaviors and origin of the known universe. As prescribed by Hubble's Law, every point in the universe is expanding away from one another at a prescribed rate; under the assumption of a Big Bang scenario, by measuring this observed rate of expansion, we can calculate how much smaller the universe was at an arbitrarily long time ago. Likewise, if we take this calculation and continue to extrapolate backwards sufficiently far into the past, we reach a point where the universe becomes so small that it calculates to be a single point – a "singularity", corresponding to a time roughly within the range of 13¾ to 14½ billion years ago. This is essentially the basis for our calculations on the "age" of the universe.[3,4,5]

One of the strongest pieces of evidence towards this model is the observed approximate 3K[†] cosmic microwave background radiation.[6] Theory holds that the source of this radiation is the afterglow left by the extremely hot origins of the universe. Analogous to how physics describes the expansion of a gas, the cosmos may be treated in much the same way – as it expands, it also cools. A gas which initially is sufficiently hot, as supposedly was the primordial universe very soon after the Big Bang, will exist in the plasma state. To this end, the excited electrons of this state would elicit poor transparency; to put it simply, the process of photons

---

[†] More precisely, the generally accepted temperature of the cosmic microwave background is $2.725 \pm 0.002K$.

being continuously exchanged between the ions would make it difficult for light to freely transverse. This would have effectively made the early universe essentially opaque to light; photons would simply bounce from one ionized atom to another, constantly being absorbed and re-emitted. It was not until the universe had expanded sufficiently and cooled enough for electrons to bind to atoms and form stable molecules, chiefly diatomic H_2, and to a lesser extent neutral mono-atomic helium, that photons had become free to transverse the vastness of space. For a dense mostly-hydrogen plasma, this occurs at temperatures below about 3,000K, where the rate of ionization of molecular hydrogen into plasma sharply decreases, and the gaseous state predominates.[7] Physicists describe this as a "photon decoupling" event occurring just as the temperature of the expanding universe drops below this critical temperature of approximately 3,000K, which correlates to a universal age of roughly 380,000 years after the hypothesized Big Bang. Some of these first photons to freely transverse the universe are still faintly visible today, if we employ the proper sensitive equipment. The thermal signature of what we observe corresponds to a temperature of 2.725 ± 0.002K, yet the spectral signature indicates a high redshift. After travelling billions of years, and subject to the expanding cosmos, we find the proper temperature (when emitted) corresponds to the deionization temperature of a dense hydrogen plasma, occurring at a time when the universe was approximately 380,000 years old compared to a current age of approximately 13¾ to 14½ billion years. It appears to be the incontrovertible, smoking-gun proof that the Big Bang model has merit.

And, it shall not be argued that the Big Bang model does not have merit, for it most certainly does; just as alchemy had merit with respect to the development of chemistry. Likewise, there remain a few problems. Starting from the beginning, one such issue is the *"inflationary epoch"*, which requires a period of greatly accelerated expansion of the universe in the fractions of a second just following the Big Bang.[8] The concept of inflationism modifies the otherwise constant rate of expansion of the universe – that is, as set forth by Hubble's constant and conceptualized as the Hubble flow – such that one arrives at a universe which exhibits the large scale structure and properties as we experimentally observe in the present day. While this solves the shortcomings of Big Bang cosmology, it is easy to see the danger in accepting this. The opaqueness of the universe prior to the recombination epoch precludes direct experimental observation of such cosmic inflation. Though mathematically sound – and despite the fact that inflationism is currently the best model we have – given the absence of observational evidence by which to validate the theory, its acceptance as a viable explanation unfortunately represents a leap of faith.

Even if we discount the probability of a Big Bang scenario, the 3K cosmic microwave background radiation unequivocally remains of great importance to our understanding of the structure and origin of the cosmos. The fact that it approximates ideal black-body radiation so well (to the extent that the word "approximates" becomes a misnomer) cannot be casually discarded.[9] If one imagines that the origin of this radiation was not the center of the universe, but at the very edge (the event horizon, the Hubble radius at a distance of r_{hs}) much of the argument towards its origins still in fact holds true. A reasonable explanation to this radiation and its observed nature could be attributed to the creation of matter across the spherical surface at r_{hs}, where, from the perspective of any hypothetical observer, virtual particles do not reunite with their counterparts. This can be conceptualized in the same manner as virtual particle pairs becoming separated across the event horizon of a black hole; the escaping particle carries away mass–energy.[10] These relativistic, fundamental particles have the capability to mimic our pretenses for the early state of the universe at the heart of the Big Bang; escaping photons across the gamut of the electromagnetic spectrum responsible for the observed black-body approximation. As previously explained, statistically protons and electrons would represent the majority of primordial subatomic particles created in the same manner, likely resulting in the formation of elemental hydrogen. Once these hydrogen atoms have cooled sufficiently, and have become transparent to electromagnetic radiation at a temperature of approximately 3,000K, they would be free to emit their thermal radiation across the cosmos. By the time it had reached us however, its measured temperature would have cooled to around a mere approximate 3K due to the cumulative effects from travelling the maximal distance through an expanding universe. It would therefore be indistinguishable had this radiation come either from the origin of the early universe, or from a distance of r_{hs} away. Paradoxically, as it turns out, both can be interpreted as exactly the same thing.

Another expected piece of evidence in support the Big Bang hypothesis which is inexplicably missing, manifests in the complete lack of any unobserved *"population III"* stars, *anywhere* in the universe.[11] These are the stars which would have been created as the first generation of stars following the Big Bang, their signature characteristic being the lack of all but the lightest of metals. (In this respect, astronomers frequently assign a loose definition to the term *metal*, typically meaning *any* element which is heavier than helium.[12]) These heavier elements have very low probabilities for being created spontaneously during a hypothesized Big Bang event, or any alternative theory which we have thus far put forth; therefore, we expect not to observe these heavier elements in any significant quantities. Furthermore, stars containing very heavy metals (notably elements heavier than bismuth) must be at least second-generation stars, as the only process capable of creating them in any appreciable quantity is under the extremes of stellar collapse – *i.e.*, nucleosynthesis in supernovae.

High-mass stars would quickly burn through their fuel in the range of millions of years to a few billion years.[13] Assuming the distribution of stellar sizes was at least remotely consistent, we should observe lower mass stars with lifetimes greater than 14½ billion years – yet there is a *complete* absence. Not one, single *population III* star has ever been observed in the entire universe. This is yet another obstacle for Big Bang cosmology. It almost certainly represents an omission or oversight is present, or else the mechanics governing stellar formation and evolution must have been dramatically different in the early to middle-aged universe than what we observe today. However, as of yet we have no evidence that the laws of the universe change over time, and in fact have quite surmountable evidence that the opposite is true. Without exception, every observation ever made in the history of science supports that the laws of physics are universal, in both time and space.

Aside from our problems reconciling the origin of *everything*, two concepts physics has yet to shed light on (for lack of a more appropriate transition) are the ill-defined entities of dark matter and dark energy. Physicists label these terms as "dark" not to allude to their malevolence, but simply to express the unknown. Dark matter refers to the apparent "missing mass" that is *supposed to* exist to satisfy the equations of general relativity in a way which matches up with astronomical observations.

If we look at a particular historic problem in physics, that being the calculation of the precession of the orbital perihelion of planet Mercury, we see that some of the solutions at the time just before the introduction of Einstein's theory of relativity made feeble and fragile mathematical "tweaks" to existing frameworks. Newtonian mechanics could approximately predict the orbit of Mercury to a remarkable degree; its orbit was understood to not be perfect due to influences, or perturbations by the other planets. However, the observed motions of the orbit could not be reconciled *exactly* – there was some unknown component to its orbit which could not be ascribed to measurement error.[14] In order to explain the discrepancy of motion, several contrived theories were put forth; some explained the deviation as the result of undiscovered mass; others introduced mathematical "fudge-factors" into proven equations. Other theories postulated even more exotic explanations still. The fact is, any one of these "solutions" possessed the capability to eloquently explain observations, and provide reasons as to why they differed from prevailing theory; unfortunately, none of these were correct.

All of this changed of course, when Einstein introduced his groundbreaking theories of special and general relativity – theories which for the first time offered a proper, rigorous explanation with the then-radical concept that space itself had properties that were subject to change. The precession of Mercury's orbit could now be explained in such a manner that did not depend on contrived notions or constants. Today, this represents possibly one of the most famous "litmus tests" of the theory of general relativity.[15]

Another currently well-known but ill-explained problem in astronomy is that known as the *galaxy rotation problem.* Nearly universally, all galaxies observed thus far appear to rotate in a manner contradictory to existing orbital mechanics. Stars located in a wide swath between the core and the edge of the galactic disc seem to orbit the galactic core at roughly the same rate. Since the masses of stars fall within a predictable range, the only other explanation using the existing framework of general relativity points to there being some unseen, unobserved mass contained within the galaxy.[16] We use the term "dark matter" to refer to this mass because it does not interact in any perceptible way with the rest of the visible universe except gravitationally, which is the only concrete evidence thus far inferred of its existence. Current models imply that objects and energy and light can freely transverse through this seemingly unknown substance without being changed; ordinary matter does not interact, and instead passes through unimpeded.

Even when we attempt to simulate dark matter using advanced computers, the results are often in disagreement with observation. For instance, our current understanding of the behaviors of dark matter, made via observations, require such dark matter to be spread out relatively smoothly throughout a galactic disc. However, when simulated using known mathematics and equations, it concentrates sharply at the center, its central density rising to a point described as a "cusp"; hence this is known to astronomers as the *core-cusp problem.*[17] This, a contradiction to the observed behavior of matter in galaxies, is a specific manifestation of the broader aforementioned galaxy rotation problem. Other problems arise with respect to the number of satellite galaxies in orbit around larger ones; when processed through the simulations, again using known laws of physics and the equations proven to work on the smaller scales of planets and individual stars, the numbers of these galaxies reported are fantastically higher. And, this should tell us something – for what is a computer simulation other than the pure mathematics and logic of a theory calculated on a large scale, unaffected by our biases? Rather than trying to find and "add in" this missing mass, using a mysterious substance, it may be more appropriate to find, to hone in on the missing piece of the theory which is responsible for these ill-explained effects.

The same could be said of "dark energy", which refers to the unknown mechanisms propelling the expansion of the universe.[18] Experiments which probe its identity do so on the very frontier of science. Dark energy is something which fits into our existing frameworks of reality *even less* than dark matter. This "energy" – which is the hypothesized impetus to the persistent and pervasive expansion of all space – has origins which evade identity at the present. Or, alternatively, it could be that we have overlooked something, and this dark energy is merely a property of space-time which is hidden in what we already know. I assume our understanding of dark energy – whatever its identity is – will rapidly change. Our relentless march forward in the understanding of the physical world in these modern times rarely tolerates the unknown for very long. Likewise, as we have begun to allude to herein, all of these mechanisms may be intimately interrelated, and a further development on general relativity which just *slightly* advances it further, may explain it all.

1   Freeman, D. "Why Revive 'Cosmos?' Neil DeGrasse Tyson Says Just About Everything We Know Has Changed". *Interview with Neil DeGrasse Tyson.* (2014). The Huffington Post.

2   Eddy, M. D. "An Introduction to Chemical Knowledge in the Early Modern World". (2015). Durham University.

3   Planck Collaboration. "Planck 2015 Results. XIII. Cosmological parameters.". (2016). Astronomy & Astrophysics, Manuscript.

4   Spergel, D.; *et. al.* "The age of the universe". (1997). National Academy of Sciences and Engineering. (Vol. 94, 6579 – 6584).

5   Alles, D. L. "The Evolution of the Universe". (2013). Western Washington University.

6   Smoot, G. F. "Cosmic microwave background radiation anisotropies: Their discovery and utilization". (2007). Reviews of Modern Physics. (Vol. 79, 1349 – 1379).

7   Langmuir, I. "The dissociation of hydrogen into atoms." (1912). Journal of the American Chemical Society. (Vol. 34, 860 – 877).

8   Riotto, A. "Inflation and the Theory of Cosmological Perturbations". (2002). National Institute for Nuclear Physics (*Istituto Nazionale di Fisica Nucleare*).

9   Wright, E. "Cosmic Background Radiation". (2001, 2006). Encyclopedia of Astronomy and Astrophysics. Taylor & Francis.

10   Hawking, S. W. "The Quantum Mechanics of Black Holes". (1977). Scientific American. (Vol. 236, 34 – 49).

11   Alighieri, S.; *et. al.* "The search for Population III stars". (2008). International Astronomical Union.

12   Frebel, A. "Reconstructing the Cosmic Evolution of the Chemical Elements". (2014). Dædalus.

13   Beuther, H. "The Formation of Massive Stars". (2006). Max Planck Institute for Astronomy.

14   Bootello, J. "Angular Precession of Elliptic Orbits. Mercury". (2012). International Journal of Astronomy and Astrophysics. (Vol. 2, 249 – 255).

15   Chandrasekhar, S. "The Role of General Relativity in Astronomy: Retrospect and Prospect". (1980). Journal of Astrophysics and Astronomy. (Vol. 1, 33 – 45).

16   Persic, M. "The universal rotation curve of spiral galaxies – The dark matter connection." (1996). Monthly Notices of the Royal Astronomical Society. (Vol. 281, 27 – 47).

17   De Blok, W. J. G. "The Core-Cusp Problem". (2009). Advances in Astrophysics, Special Issue.

18   "Understanding Dark Energy". (August 2015). United States Department of Energy, Fermi National Accelerator Laboratory (Fermilab).

VII

WATCHING OUT FOR FALLING OBJECTS

IT MAY BE SAID, that everything in the world around us is *falling*. On Earth, if we drop a pebble – or a paperweight, or Newton's apple, or *anything* for that matter – it falls towards the ground. Essentially, everything around us wound up here because, at some point or another, that is where it fell. This realization is nothing new – mankind has tirelessly pondered over this concept of bizarre attraction for thousands of years. The great ancient Greek philosopher Aristotle, long ago in antiquity, postulated that it is the very nature of particular materials to "gravitate" towards their natural place. Through observation and logic, he reasoned that heavy materials such as stone and lead sought to be closest to the center of the Earth – then thought to be the center of the universe at that particular time. The "material" of fire, on the other hand, sought to be closest to the heavens.[1,2] Such was the theory of what we today call gravity, as it persisted for millennia until Newton was inspired, supposedly, by the infamous falling fruit. The "*eureka!*" moment came of course when Newton reasoned, just as the apple falls toward the ground, so does the Moon.

Extending this further, the Earth is falling around the sun; and, in turn, the sun falls around the galactic center. Small galaxies fall around larger ones. Even the energy released by stars undergoing nuclear fusion is done so by atoms falling around a stellar core – which, again, accreted in the first place by things falling onto it. Those atoms which eventually fuse do so under the weight and pressure of things falling on top of it. One contrarily may point out the fact that opposite charges attract each other at a separation distance, independent of where things fall; and this is true, there are other forces besides gravity. However, on large scales of organization, none seem to influence the cosmos so profoundly.

The overwhelming majority of matter in the universe consists of hydrogen and subatomic particles smaller than hydrogen, such as protons and electrons.[3] There are a sleuth of other subatomic particles which are inherently unstable and thus short lived – the result of collisions between mostly free electrons, protons, and combinations thereof. But these can be regarded as transient states, as ultimately they decay into stable protons or electrons, or their antiparticles – or alternatively, they may decay into electromagnetic radiation as they cease to exist as matter at all. Whatever the origin of matter, whether from a hot Big Bang or otherwise, it seems to favor the production of matter which is predominantly stable; unstable matter tends to decay into ionizing radiation and would be a threat to life as we know it.

The universe is not expected to bear mass within the confines of an isolated system – otherwise, the entire idea of conservation of mass-energy would become moot. One way that the mysterious origins of everything can be reconciled rests in the case where it is supposed that all of the matter and energy contained within the visible universe was manifested either very long ago, or very far away. If we say that each approaches infinity – that is approaching infinitely long ago in time, and approaching infinitely far away in distance – we may find that, strangely, both really mean sort of the same thing. A distance of billions of light years towards the horizons of the cosmos also implies a time billions of years ago.

The question should be raised as to why the leading cosmological model seems to illogically place all of everything inside an infinitesimal space, only to depend on everything trying to come back together again. This leaves another enigma as to where this enormous energy came from, to fling all of the universe apart. It is worth entertaining that the "birth" of the universe may not have been so violent; alternatively, there may be a more elegant solution.

Imagine, for a moment, the nature of virtual particle pair production at a distance of the Hubble radius, r_{hs}, from an observer located at some hypothetical fixed point. From that vantage point, such an event would be occurring at precisely the age of the visible universe, and at the maximum possible distance from that observer. Beyond the distance r_{hs} represents that portion of the universe which is separated by an event horizon, impervious to the transmittal of information. Just as the previously discussed case of a black hole, virtual particle pairs which are separated by such an event horizon, are destined to "fall into" the universe, thereby giving rise to real mass and energy.

Both observation of the local universe in the present day, and the much earlier universe via the cosmic background radiation, seem to suggest that the dominant constituents of primordial matter in the universe were the lighter elements of hydrogen, and to a lesser extent elemental helium. As a matter of probabilities, this makes sense. If the creation of massive particles depends on virtual particle production across an event horizon, such a process would statistically favor the production of the most basic, stable forms of matter. Anything else would either be too unstable and quickly decay, or would be improbable and thus increasingly rare with increasing complexity.

Simple protium, ^1H, and its sole complementary electron comprise the most fundamental and stable common constituent of atomic matter in the observable universe. Neutrons may also appear at first glance to be a logical choice, however free neutrons decay quickly – the half-life of a free neutron being a scant approximate fifteen minutes – and thus make them an unlikely candidates to exist unbound.[4] Such serves to strengthen the argument for this previously described distant pair production hypothesis.

So then, as a theoretically infinite gamut of vacuum fluctuations inevitably lose their virtual characteristics when separated across the event horizon at r_{hs}, most particles ultimately arising from the available energy will be unstable and short lived, merely due to statistics. However, a small number of these fluctuations will produce long-lived varieties of particles such as protons and electrons – either by producing constituent particles of the proton and/or the electron – if any yet unknown constituents exist for the electron – or by producing the whole particles themselves. We are also aware that even many unstable particles exhibit decay modes which often ultimately arrive at protons and electrons anyway.

Therefore, it stands to reason that the most common massive particles generated across the event horizon at r_{hs} would likely be protons and electrons. As probabilistically the most common primordial products of genesis, observation likewise supports this. Allowing these primordial particles to cool sufficiently – below approximately 3,000K, or the first ionization energy of H_2 – these protons could and electrons would be able to combine into stable hydrogen. If the cooling was due exclusively, or almost exclusively, to the expansion of the universe, it is expected that this would occur after roughly 380,000 years.[5] We see this described within Big Bang theories as the *"recombination epoch"*; understandably, the distinction of whether this occurred a long time ago, or simply far away, is not immediately obvious under either paradigm.

After combining into stable hydrogen, the predominant force governing these particles would be gravity. Any small perturbation – even so small as being struck by a stray photon – would ever-so-slightly nudge a single hydrogen atom or molecule of H_2, so that it was just ever-so-closer to the next. Such would thereby effectively introduce a degree of anisotropy, creating a local concentration of mass. Over time, even only the slightest nudges of the hydrogen from any possible isotropy, would cause them to condense into clouds, and when sufficiently massive, stars.[6] As one may surmise, these stars would be apt to coalesce into clusters, and then ultimately galaxies and so forth into galaxy groups and filaments. Such would be a process afforded not only of the past, but a continual one.

This somewhat in contrast to what we are led to believe via Big Bang cosmology, where it is described that all of existence "exploded" from a single point – yet resulted in an end-state universe whereby the predominant force is gravity. Under varying rates of expansion, as explained in inflationism, a pre-existing "singularity" was ripped apart to yield the current, expanding universe we observe today. It seems such a model needs many necessary layers of explaining – *e.g.*, inflationism, dark energy, dark matter, *etc.* - whereby the aforementioned alternative seems to avoid all of these unnecessary modifications.

To this end, via this process one would expect to see more "primitive" galaxies forming at great distances – for it is at these great distances we are looking at object very long ago and very close the event horizon at a distance of r_{hs}.[7,8] Such an observation might support the Big Bang hypothesis in the sense that we are observing object which are very young; yet it should not be construed that these observations are of a time when the universe *as a whole* was young. Furthermore, our conceptions of age are only accurate if there is some *absolute* age to the universe. If this is not the case, then our age estimates are meaningless – they only tell of some time difference from the present. The only thing we can claim with near certainty, is quantification of *relative* times – from our vantage point – describing how distant objects looked in the *relative* past.

1   Papaspirou, P.; Moussas, X. "A Brief Tour into the History of Gravity: from Democritus to Einstein". (2013). American Journal of Space Science (Vol. 1, 33 – 45).

2   Mittelstraß, J. "The Concept of Nature – from Plato's World to Einstein's World". (2014). The Pontifical Academy of Sciences, Plenary Session on Evolving Concepts of Nature.

3   Weinberg, S. "The First Three Minutes: A modern view if the origin of the universe". (1979).

4   Heilbronn, L. "Neutron Properties and Definitions". Supplement. (2015). Nation Aeronautics and Space Administration (NASA), Johnson Space Center.

5   Smoot, G. F. "Cosmic microwave background radiation anisotropies: Their discovery and utilization". (2007). Reviews of Modern Physics. (Vol. 79, 1349 – 1379).

6   Bromm, V., et. al. "The formation of the first stars and galaxies". (2009). Nature (Vol. 459, 49 – 54).

7   Oesch, P. A.; et. al. "A remarkably luminous galaxy at z = 11.1 measured with Hubble Space Telescope grism spectroscopy". (2016). Astrophysics Journal. (Vol. 819, No. 2).

8   Murdin, P. "Galaxies at High Redshift". Encyclopedia of Astronomy & Astrophysics. (2006).

– VIII –

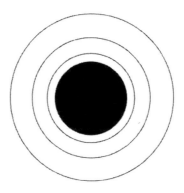

Interesting Horizons

IN THE CHILLY EARLY MONTHS OF 1916, a physicist by the name of Karl Schwarzschild provided a solution to Einstein's recently published field equations describing the distortion of space-time around a massive object.[1] Serving in the armed forces during the first World War, Schwarzschild wrote a letter to Einstein, who was quite impressed with his handiwork. Unfortunately, shortly thereafter Schwarzschild died near the turmoil of the front lines, and consequently, this first solution to the field equations was his only. However, Schwarzschild's solution has turned out to be especially important. One aspect gives us a way to describe massive objects – really massive objects – especially dead stellar cores such as neutron stars and black holes. It is in fact his name that several parameters and equations bear to this day, still composing, in part, a few aspects of our contemporary fundamental understanding of the subject.

Specifically, Schwarzschild's solution describes the space-time surrounding a static, spherical, non-rotating massive object. While the solution as a whole has numerous applications, let us consider one of the derived equations:

$$R_S = \frac{2GM}{c^2}$$

Here R_s is contemporarily known as the "*Schwarzschild radius*" of a described object with mass M. The term denoted with large G once again assumes the role of the universal gravitational constant. R_s denotes the closest distance beyond which the escape speed from the object with mass M exceeds the speed of light, c. This equation describes an imaginary spherical surface at distance R_s from the center of the object with mass M across which the exchange of information is limited to one direction – in other words, an "*event horizon*".

One example of an event horizon (and probably the most familiar) exists surrounding a black hole. This is the "*point of no return*", the imaginary spherical surface surrounding the black hole – although some distance away from the object itself – which can be crossed in the direction toward the singularity, but once crossed, one cannot travel back across it. This is one example of an event horizon, which occurs *anywhere* space-time has been distorted to confer an approach or recession above the speed of light. Hubble's constant provides for an expansion of space-time which is proportional to distance; that is the greater the distance considered over a period of time, the more extreme the change in relative velocity between two distant points becomes. At a certain threshold, any two points unbound by the fundamental forces (*e.g.*, gravity, electromagnetism, *etc.*) will be receding from each other at greater than the speed of light, and an event horizon will effectively be formed between them against the direction of expansion. This radius becomes none other than the Hubble length, or r_{hs}. Therefore, we may state that $R_s = r_{hs}$.

We can illustrate such is true via the following, simply by canceling like terms in the following expression:

$$R_s = \frac{2GM}{c^2} = \frac{2G}{c^2} \times \rho_c \times V_{hs}$$

$$= \frac{2G}{c^2} \times \frac{3H_0^2}{8\pi G} \times \frac{4}{3}\pi \times \left(\frac{c}{H_0}\right)^3 = \frac{c}{H_0}$$

Again, the *M* term in our above calculation is the mass of a typical Hubble sphere, as given earlier by:

$$M = \rho_c \times V_{hs} = \frac{3H_0^2}{8\pi G} \times \frac{4}{3}\pi \left(\frac{c}{H_0}\right)^3$$

The mathematics of Schwarzschild's solution are quite sound. As one of the earliest solutions, it has been validated time and time again – including by Albert Einstein himself – in numerous contexts and applications. Impressed, he eagerly wrote Schwarzschild in response to his solution:

> "I have read your paper with the utmost interest. I had not expected that one could formulate the exact solution of the problem in such a simple way. I liked very much your mathematical treatment of the subject. Next Thursday I shall present the work to the Academy with a few words of explanation."
>
> – Albert Einstein,
> correspondence to Schwarzschild, *circa* 1916

Since the time of Einstein's correspondence, we have surmounted strong and convincing experimental and observational evidence to validate the predictions of this solution.[2] Additionally, by extension, his calculations have even formed the basis for new theoretical work. It is rather a shame that Schwarzschild's time on the subject of general relativity was so very brief. It certainly begs the question of what other interesting solutions could have been conceived, had Schwarzschild been afforded the luxury of more time. Likewise it is a misfortune that he did not live long enough to witness the irrefutable proof of general relativity just a few years later.

In 1919, physicist and astronomer Sir Arthur Eddington traveled to the island of *Príncipe* off the coast of Africa to capture images of the total solar eclipse of May 29th in that year.[3] It was in those photographs taken of the eclipsed sun, where a few stars which should have been obscured *behind* the sun, were actually instead seen *beside* the sun. The light from these distant stars, as a result of the enormous gravity of the sun, was actually being bent through the gradient density of the compressed surrounding space – effectively in much in the same way light is refracted through a glass lens. Overnight, humanity's understanding of the universe was profoundly changed, and Einstein's name was well on its way to becoming one which would not soon be forgotten; and, by extension, the legacy of Karl Schwarzschild became inextricably intertwined as a pillar of our modern understanding of the cosmos.

Perhaps another one of the most incredible discoveries of the twentieth century is that empty space is anything but empty. What classical mechanics considers the *lack* of all things, a region of space devoid of everything, turns out to actually be a very real, identifiable medium with measurable properties susceptible to change, analogous to, but wholly different than those of everyday matter. This is not merely some clever abstract achievement of physics; it has profound implications. These properties of empty space govern the behaviors and consequences of the matter and energy contained within. The field of quantum mechanics – a subject area from which only small elements will be given at best a cursory overview herein – deals with the interaction of matter and energy in space-time at the most miniscule of scales. As oft-quoted, "*If quantum mechanics hasn't profoundly shocked you*", famed physicist Niels Bohr once said, "*you haven't understood it yet.*" Despite its apparent weirdness and perceived lack of intuitiveness (initially at least), quantum mechanics does an exceedingly elegant job of mathematically describing these interactions. We are presented with one of nature's most beautiful enigmas; to understand the very large, we must first understand the very tiny.

Unfortunately this does nothing to counter the attestation that the effects of quantum mechanics are nothing short of baffling by our familiar macroscopic standards. We are accustomed to observing the world on large scales, and what we see – what we perceive as realities – are generalizations of vast swaths of quantum mechanical interactions. We rely on the statistical likelihood that these enormous sums of interactions to even the smallest events which are relevant to us will manifest in similar, or even identical ways, time after time. This is the basis for the innate *intuition* of truth that we possess, despite the fact that as the scales become smaller and smaller, this intuition ends up being progressively incorrect.

Therefore, of all the strangeness that occurs in the vicinity of and across an event horizon, here we are most interested in the smallest of things. Theoretical physicist and mathematician Stephen W. Hawking has proposed mechanisms by which, due to quantum effects, mass and energy are able to transverse an event horizon, effectively taking mass away from term M. To give a very nontechnical explanation, vacuum fluctuations in the form of virtual particle–antiparticle pairs are continuously formed an annihilated over very short distances and very brief times.[†] These short-lived reactions facilitate a number of quantum phenomena; in the case where these pairs are produced in close proximity to an event horizon, it is possible for one to transverse the event horizon and thus never be able to meet back with its complement. The "free" particle is able to "escape", and via the laws of conservation, it mass-energy is lost from the opposite side of the event horizon – a normally forbidden feat.[4]

In Hawking's book, "*The Universe in a Nutshell*", he shows a useful diagram for conceptualizing the processes of virtual particle pair production and subsequent annihilation, and would I urge anyone who is interested to have a look for themselves. Hawking, after

---

[†] One must be careful to note that the term *antiparticle*, in the context of virtual particles, is *not* the same as when we talk about particles of antimatter versus normal matter; antiparticles in the virtual sense describe something fundamentally different.

all, a progenitor of this theory enabling particles to cross an event horizon in this manner in the first place. In the meantime however, perhaps the schematic, opposite, will suffice. Here, in this rudimentary two-dimensional diagram, we have depicted two pairs of produced virtual particles from vacuum fluctuations of empty space. The large circle labeled r_{hs} represents a hypothetical event horizon. At point A. a typical pair is produced with its antiparticle, and the two proceed to annihilate each other as expected at point B. The trajectories from point A. to point B. represent the vast majority of vacuum fluctuations, where nothing particularly interesting is occurring. The virtual particle pair produced at point C. is a special case which we have just discussed. Its proximity to the event horizon r_{hs} causes one virtual particle to transverse the event horizon, and arrive a point such as D., never to interact with the outside world again.

This finally leads us to the trajectory of interest, the particle which ends up at point E. This particle, without a virtual antiparticle complement, is now traveling outside the event horizon, and becomes indistinguishable in its properties from any other particle in the universe. Due to the laws of conservation, it *must* carry away mass-energy from *within* the event horizon. In the typical case where r_{hs} represents the event horizon surrounding a black hole, it results in the black hole loosing mass. This is analogous to the black hole "evaporating", as the mechanism of the theory is

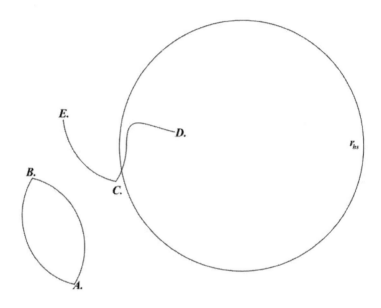

Schematic depicting mass-energy genesis via crossing
an event horizon by way of virtual particles.

sometimes popularly described.

At first glance it may appear to be an unintuitive explanation for how mass gets carried away from inside the event horizon, as the particle pair is created *outside* and then one falls in, the other escaping. It also appears that two particles have been created and the conservation of mass–energy has been violated; this is not the case. Energy is conserved, in that one particle has *positive* energy, while the other particle has the exact amount of *negative* energy; thus when they contact and annihilate, the net result is zero. The most intuitive conceptualization of negative energy would be to think of it implying that such a virtual particle "borrows" energy from the future in order to exist. In a sense, it exists in the past, and travels backwards though time – a behavior that is "legal" within the framework of quantum mechanics for very small things over very short times – *e.g.*, an electron or a photon over the course of extraordinarily brief durations. With larger or more complicated objects, and longer lengths of time, the probability of this occurring falls off so rapidly, that it is simply not observed. Again, this is a very casual and nontechnical explanation of the weirdness of quantum mechanics, without diverging too far from the scope of topic at hand.

Hawking's work on this subject has been chiefly on the study of black holes, and how they lose mass over time – in other words as we have said before they "evaporate" in a sense. There are other potential applications; while it may appear to be a specialized process, nothing states this mechanism is exclusive to the evaporation of black holes. Since this happens *because* there is an event horizon, the same concept can be extended *anywhere* there is an event horizon. Within our universe – as shown in the preceding diagram – at any given distance r_{hs} from an observer, we have an event horizon.

Many times, when we think of physicists talking about black holes and event horizons, images of them pulling out diagrams with lots of wavy lines come to mind; or we end up with references to light cones and other confusing constructs; sometimes they even speak of large rubber sheets, and heavy bowling balls. Whatever the analogy employed, realize that they are just that – analogies, a symbolic comparison; an attempt at simplifying a more complicated concept. More often than they should be, these analogies are confusing, or misleading – even blatantly incorrect, occasionally. The worst offender (in terms of popularity) is the ubiquitous rubber sheet, which seems to pervasively invade every respectable physics book published within the last hundred years. The topography has merit, but the *stretching* is misleading; in the vicinity of a massive object, theory conversely describes how space-time will *compress*. This is an important distinction, because an event horizon will form in *either* scenario – that is, an event horizon occurs when space-time is either compressed past a certain degree, *or* stretched to a similar degree. The difference between them is in the directionality which information may be passed – event horizons always result in a *one-way* restriction of the passage of information.

Let us consider the diagram below. Here we are examining the curvature of space along a line extending from an extremely massive object, located at point M., towards the event horizon of the visible universe, from the perspective of an observer in the relatively flat region denoted *s*.

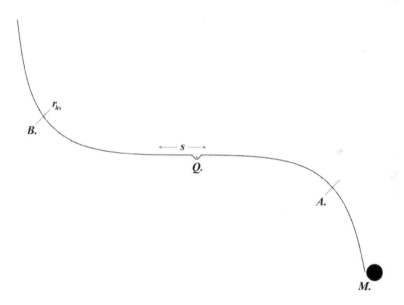

Compression of space-time is depicted with a downward deflection, expansion by an upward deflection, while flat space-time maintains the straight and level line. An event horizon is seen at location A., caused by the distortions of the sufficiently massive object M. – *i.e.*, a black hole. The event horizon depicted at location B. is that at distance r_{hs} away from a hypothetical observer somewhere in region s, and is a result of the stretching of space-time due to the pervasive expansion of the universe. The small distortion, or "gravity well" seen at location Q. represents a typical massive object, such as a planet or a star or a galaxy, which appreciably distorts space-time via compression, but lacks sufficient mass to form an event horizon. Information is free to transverse either horizon in the general outward direction away from region s towards either mass M. or through r_{hs}, but not the other way around – that is, neither *into* the observable universe, nor *out of* the black hole.

Let us consider for a moment some of the implications of an expanding universe near the event horizon r_{hs} from an observer centered within V_{hs} – recalling that V_{hs} is the "*Hubble sphere*" containing the visible universe from the perspective of a central observer. The continual expansion of space between the observer and an arbitrarily distant object (at some remote point) will eventually cause that object to recede from the observer at an ever increasing rate. Eventually, this hypothetical object – which may be a star, a galaxy, or any other astronomical body – will appear to recede at a relative velocity which exceeds the speed of light. Remember, we are not stating that the object is traveling through space at a speed which exceeds that of light, but rather it is the effect of the expansion of space between the observer and the object. We may instead think of it as the space which *contains* the distant object is moving away from the space which contains the observer at a rate which exceeds the speed of light. It is at precisely this co-moving distance V_{hs} where this transition occurs, that we find the event horizon of the visible universe.

Continuing with our thought experiment, let us now instantaneously trade places with our imaginary object, where we are now at the co-moving distance r_{hs}, and our imaginary object is at a distance of $-r_{hs}$, or the same distance in the opposite direction away from us. If we were to look around our new cosmological neighborhood, we would notice that we can now only see somewhat less than half of the objects within the universe that we were previously able to observe.

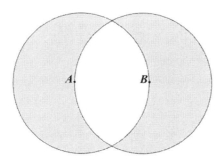

We can represent this with a *Venn* diagram, as the intersection of two different Hubble spheres with their centers separated by exactly r_{hs}. In this rendition, our original location would have been at point A., with our transposed position now at point B. We will also notice, towards the direction we have moved our perspective, we can now see an additional distance r_{hs}. This also implies that we can see events occurring at an interval of H_0^{-1} in the past. As we have explained earlier, H_0^{-1}, or *Hubble time*, forms our best current age estimate of the universe. While this poses no issues to the observer who has not moved, it is apparent that motion of any distance $d \cdot r_{hs}^{-1}$ (that is, distance d divided by r_{hs}) will then have the effect of allowing that observer to view events $H_0^{-1} - (d \cdot r_{hs}^{-1})$ in the past. In other words, that observer will now see events occurring $d \cdot r_{hs}^{-1}$ seconds *before* the current estimated age of the universe. While this may be the case however, any attempt for the observer to measure the age of the universe from his new position will fail to yield an older-appearing universe. Assuming Hubble's constant is indeed constant, that observer will *still* measure the universe to be its approximate fourteen billion years old estimate that our measurements would also agree with. Consequently, the continuity of measured apparent age extends not only to spans of distance, but also spans of time; since r_{hs} remains constant, this requires H_0^{-1} to remain constant as well. Despite these peculiarities, such are the intuitive conclusions which can be drawn upon the model of a continual Hubble flow. The notion of the universe having a beginning becomes an argument which no longer makes sense. This is a bold revelation, one which disagrees with the existence of the Big Bang. Absolute bounds to the universe equivalently make no sense; all time, and all position all become relative.

From the classical perspective, it would appear that matter had originated from inside the nothingness of a point at the origin of the Big Bang. To understand what we have just previously described, we need to take this conceptualization and spatially turn it "inside out". Surprisingly, doing so still remarkably preserves long-standing theoretical frameworks. The universe still *appears* roughly fourteen billion years old; and the universe is expanding as just the experimentally measured rate. Peculiarly enough, *all* matter still springs into existence roughly over an interval of approximately fourteen billion years ago, from our frame of reference. This is an important similarity, because it shows not that Big Bang cosmology is horribly wrong, but simply that it has made a relatively minor error in conceptualization. In other words, the mathematics of the theory are sound, the methodology we use to experimentally test the theory are equally sound. It is primarily in the way the theory is mentally visualized and interpreted that exhibits the flaw.

Perhaps up until about a hundred or so years ago, the prevailing theory many astronomers held described the universe as flat, static, and infinite.[5] The further they could peer with their increasingly powerful telescopes, the more stars and "nebulae" they could see[†]. It wasn't

---

[†] Historically, stellar nebula and galaxies were referred to under the same umbrella term of "nebulae" before it was realized galaxies were themselves actually vast collections of stars.

until the epiphany of "island universes", and the discoveries of Edwin Hubble in the 1920's which introduced the concept that the universe is *expanding* away in *all* directions, that the stirrings of a Big Bang theory gained momentum and credibility. It stood to logically reason, if presently all of the universe was expanding away, at some point in the past everything must have been much closer. Further extending this logic to some very distant point in the past, everything must have been *very* close, and if sufficiently far into the past, the universe must have even been point-like. This well-promulgated ideology has swayed us in our interpretations of observations in the physical world ever since.

Big Bangs are not the only known process to which the universe may give rise to mass–energy out of seemingly nothing. However, thus far from what scientists have observed, it seems virtual particle pair production (*i.e.*, vacuum fluctuations) is the only candidate process which is relatively well understood. The Casimir effect (so-named after physicist Hendrik Casimir) is one such well-known and experimentally verified process.[6] The phenomenon describes the attractive force experienced by two closely spaced conducting surfaces, as they exclude between them certain vacuum fluctuation frequencies from the infinite gamut of quantized possibilities. Due to the quantized nature of waves, only certain frequencies can exist inside the separation, while no such restrictions occur outside; the result is a measurable, net-positive force which drives the plates together.

The event horizon of the visible universe may present another example of vacuum fluctuations giving rise to mass–energy. Presented below is a diagram featuring two large concentric circles, centered about an imaginary, typical observer at an arbitrary position, point O., at any hypothetical point in space:

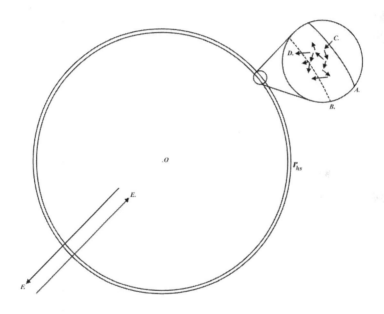

The outer circle represents the event horizon of the visible universe at a distance of r_{hs} from an observer at point O., bounding the entire visible universe. (Such bounds in actuality describe a spherical surface centered about a hypothetical observer). The magnified region in the upper right is where we shall focus our attention first. The outermost border A. represents the outer circle laying at distance r_{hs} from the observer, and constitutes an event horizon. The path denoted C. represents a virtual particle which has crossed the event horizon A., for which its complementary virtual antiparticle (not depicted) has not, and thus has become real with respect to the reference frame of our centrally located observer. This can be conceptualized as analogous to the virtual particle which "escapes" a black hole as its complementary antiparticle is trapped behind the event horizon, towards the central singularity. (As illustrated previously, this particular event horizon has information flowing outwardly in the opposite direction.) The many small arrows which subsequently follow represent the numerous collisions and combinations which occur in proximity to the event horizon as the newly created, relativistic particles interact. The dashed line within the magnified region denoted B. represents the distance of approximately H_0^{-1} – approximately 380,000 light years from the imaginary hypothetical observer at point O. This is the distance at which the expansion of the universe has sufficiently cooled the relativistic gas produced by the isolated virtual particles, akin to the time under Big Bang cosmology where the universe became transparent to electromagnetic radiation. The reason for including this in our illustration is to show how a time under one theory, can actually be interpreted as a distance in another – yet both are observations of essentially the same thing from differing vantage points.

Finally, the arrow labeled $E.$ represents the continual inward flow of isolated virtual particles (akin to many particles like $C.$) across the event horizon which move in a general way towards an observer; conversely the arrow labeled $F.$ represents the continual outward flow of matter and energy due to the perpetual expansion of space-time. It would not be surprising to find, if some method could be so devised, that the total inflow ($E.$ at all points on the surface at r_{hs}) would exactly equal the total outflow ($F.$ at all points on the surface at r_{hs}). It only seems logical that such the case would have to be true if the conservation laws of mass–energy were to be upheld. Thus far in our studies of the physical world, we have yet to encounter any evidence which points to the contrary.

Continuing along with this line of thought, everything in the entire universe, would merely be a fleeting fluctuation existing in some transient state for an unspecified duration of time. We use the term *fleeting* in the sense that ultimately a particle which has been produced as a consequence across an event horizon to a specific reference frame, our observer in the previous depiction, will, as a statistical eventuality, leave this reference frame and cease to be relevant to anything within. We could assign a value to each fluctuation – say, arbitrarily *positive* one (+1) for each virtual particle which crosses inwardly and becomes "real", and *negative* one (–1) for each particle crossing outwardly and leaving the observable universe. The outcome is essentially the same for these two particles, as would be for any other two particles created and annihilated wholly within the volume of space bounded by r_{hs}. If influx does indeed equals outflux, which would be the expected case if the conservation laws are to be observed – that is to say if all of the same types of particles are realized across the event horizon as pass outward, then the net result over long enough periods of time is zero; nothingness. Hypothetically, we could state this using the following prototypical expression:

$$-n_{x_{out}} + n_{x_{in}} = 0$$

Here, the term n_x, denoted with the "*in*" subscript is representative of the number of archetypal identical particles flowing into the observable universe – *i.e.*, positive flow. Likewise, n_x as denoted with the "*out*" subscript, represents the number of identical particles leaving the observable universe – *i.e.*, the negative flow.

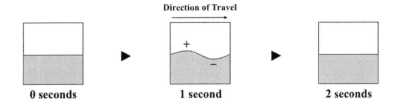

What we are implying in this explanation is that all of reality is only perceived as *something* only because we observe these sequential cross-sections of the universe *in time*, each fluctuation like snapshots of the passing of a single, solitary ripple across the surface of a pristinely calm pond.

By definition, *any* virtual particle, until it reunites with its antiparticle counterpart, can be treated as *real*; it is usually that these creation–annihilation events occur over such brief times and miniscule distances that they barely interact with anything, if anything at all. This raises the question whether or not *all* particles belong to virtual particle–antiparticle pairs over long enough times and distances. While admittedly hypothetical, if this is true, then all of the universe we observe would merely equate to a transient state, of symmetrical complementary perturbations from the ground state like waves through a medium, manifesting as reality via some interference pattern induced by the fundamental rules which everything in the universe abides. We hold the concept of matter, as in the "stuff" all material things are made of, with special reverence; and, this is not surprising, as we are made of matter ourselves, as are our family, friends, and everything else we hold dear. However, matter may merely be a quality, a manifestation, of space-time, via its structure of peculiar virtual particle pair creation and annihilation events. If this is true, what we see and known and touch and feel may be analogous to ripples on the surface of a pond, or sound waves in air – or something akin in concept to the seemingly sophisticated self-interference patterns of light – another example of a wave within a medium – which confers to produce an optical hologram. This is a curious notion, though one which has nevertheless been postulated before. Is all of existence the transient result of vacuum fluctuations? Herein presents a new perspective for an old question.

Far away from earth, astronomers have been for some time observing super-luminous objects known as *quasars*. These objects are so bright that when resolved in the early to mid-twentieth century, the luminosity of their cores completely overpowered and obscured their more complex nature.[7] Only more recently have we been able to dispel the notion that these objects are point-like sources of light. The term "quasar" derives from initial descriptions of these objects as "quasi-stellar" light sources, appearing akin to stars from a vantage point on Earth. Contemporarily, we regard quasars as objects which are likely very distant galaxies with hyperactive galactic cores. For some time before the first known quasars were cataloged, entities known as *Seyfert galaxies* (so-named after the astronomer Carl Seyfert) were observed as early as the beginning of the twentieth century. These galaxies, today described as "active" galaxies, contain very luminous central regions. The difference between these two objects likely comes down to a matter of distance; Seyfert galaxies were most likely

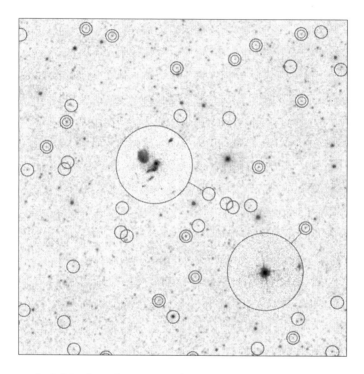

Wide-field Infrared Survey Explorer Image, Courtesy of NASA
Inset photographs contain Hubble Space Telescope imagery.

Shown here are actual telescope images of quasars, with the
main image taken by the WISE orbiting observatory, supplanted
by enlarged inset details taken with the Hubble Space Telescope.[8]

identifiable as "quasars" in close enough proximity to us that we were afforded the ability to discern other features.

It is much the same concept with stars in regard to their planets, or solar flares and sunspots. Save for the sun, we cannot directly observe these features for the same reasons; the star itself outshines these smaller details. However, in the case of distant quasars, the smaller "details" end up being the stars themselves. Depicted on the previous page is actual space telescope imagery of these objects; individual quasars are encircled within insets containing observations made by the Hubble Space Telescope.

Within these distant, active galactic cores – as the primary source of their intense illumination – are almost certainly supermassive black holes similar to the ones found at the central regions of most nearby observed galaxies. The important distinction is that they are actively accreting – that is, gas, dust, and perhaps even whole stars are falling into them. Our own galaxy possesses such an object, named Sagittarius A* (the latter pronounced as "A-*star*").[9] Despite its heft, at more than four *million* solar masses, it is relatively quiet in comparison to these distant galactic cores. They are statistically different from the other galaxies we see nearby – differing in that we see greater numbers of them at great distances. Astronomers and cosmologists have interpreted this observation to mean that these quasars are a feature of the early universe – that they were formed and existed long ago (by factors present only long ago) and are more recently a less common occurrence. Interpreted in such a manner, this becomes a strong piece of evidence in support of the Big Bang hypothesis; and, it would be convincing had there been no other way to explain their manifestation in this way. Combined with observance of the approximate 3K cosmic background radiation, it comprises one of the supporting pillars of the theory.

Of course, there is always an alternative explanation. We have already considered that the origin of matter *was* not a point-like "explosion" roughly 13¾ to 14½ billion years ago, but rather *is* a consequence quantum effects at a distant event horizon. This is another situation where we need not "throw the baby out with the bath water", so to speak. The conditions theorized to exist at the time of the Big Bang have remarkably paralleled similarities to the conditions at great distances, in proximity to the event horizon lying arbitrarily at a distance of r_{hs}.[†] If we follow a line extending *from* this distant event horizon to ourselves (the observer) we would notice quite a few similarities through how the *distance* in light years corresponds to *time* in Big Bang cosmology. Thus, the higher concentration of quasars is perceived as long ago, but more importantly because these objects are far away from us. These distant objects

---

[†] One poignant example consists of the previously depicted, approximate 380,000 light-year-deep region which can be treated as analogous to the early opaque universe under the Big Bang model.

exist "closer" to the event horizon of the visible universe, as seen from our vantage point.

What we are getting to, is that to say we are observing the *very early* universe – in the way the Big Bang model describes it – is actually only a consequence of our viewpoint. To observe an object a billion light years away, is to peer some billion years into the past. At the same time, such a distant object has had the same billion years *less* time to interact with us than objects in our immediate vicinity. Concisely, to observe an object approaching the distance r_{hs}, the number of interactions from our perspective approaches zero – that is to say fewer and fewer photons reach us, and thereby less information reaches us, the closer we approach to this distance limit. If we could imagine traveling on a ray of light to one of these galaxies, we may find upon arrival each is not much different (on large, archetypal scales) than our own Milky Way and the local galactic neighborhood. However, by the time we had reached one of these quasars, or distant Seyfert galaxies, such would be many billions of years older than when we first observed it. On the same token, looking back upon the Milky Way, along our journey it would seems to have at some point vanished, quietly drifting across the event horizon behind us. Yet, in its place, we may find a similar "young" quasar at the edge of the our new *observable* universe; such an object would appear to be a new, infantile galaxy. To ride a light ray back to this now distant object would yield a similar outcome, and so-on. The farther we look back, the more we are observing behaviors more constrained to short periods of time, in the distant past, near the event horizon of the visible cosmos. Thus is the peculiar weirdness of a flat, expanding universe.

Similarly, there is evidence in contemporary observations of a few ordinary galaxies which exist at distances near, or even beyond these quasars; we are using the term *ordinary* to describe the mature, developed galaxies like our own, which have relatively quiet central supermassive black holes.[10] While seemingly in opposition to the Big Bang hypothesis, the appearance of these ordinary galaxies at these great distances is an expected consequence of a universe demonstrating an infinitely expanding Hubble flow, where despite the overall outward flux of galaxies across the event horizon of the visible universe, there is a mixture of relative motion, to and fro from the perspective of an arbitrarily placed observer, within the galaxies. Under this reasoning, it is merely less likely to find many "old" appearing, mature galaxies at great distances – yet still statistically afforded in small numbers. This is in accordance with observation – as we build increasingly powerful telescopes, we seem to glimpse organized galaxies increasingly closer to the hypothesized Big Bang event, continually shaking our conceptions of galactic formation. Mature galaxies which feature low rates of stellar genesis have been observed at redshifts which suggests their formation within just hundreds of millions of years after the hypothesized Big Bang.[11]

Objects in space, on grand scales, behave in a manner similar to atoms in an expanding gas.[12] The motions of stars and galaxies appear random at small scales, varied relative to each other. However, if we can step back sufficiently far, we notice a different relationship; we find that on average all of the atoms are moving away from each other, roughly uniformly. Looking at a single galaxy in the universe is comparable to looking at only a single atom in an expanding gaseous nebula. Though the Milky Way and the great Andromeda galaxy are hurtling towards each other, while most other galaxies are doing precisely the opposite.[13] We would expect to find the occasional "lingering" galaxy, which for reasons unknowable save to its tortured past, have kept the image of its ghost within our view; the only evidence from long ago of its meandering ways.

1   Heinickle, C.; Hehl, F. W. "Schwarzschild and Kerr Solutions of Einstein's Field Equation". (2015). International Journal of Modern Physics D. (Vol. 24).

2   Collier, P. "A Most Incomprehensible Thing: Notes Towards a Very Gentle Introduction to the Mathematics of Relativity". (2014).

3   Coles, P. "Einstein, Eddington and the 1919 Eclipse". (2001). Astronomical Society of the Pacific Conference Proceedings. (Vol. 252).

4   Hawking, S. W. "The Quantum Mechanics of Black Holes". (1977). Scientific American. (Vol. 236, 34 – 49).

5   Kinney, W. H. "Cosmology, inflation, and the physics of nothing". (2003, 2004). Lectures given at the NATO Advanced Study Institute on Techniques and Concepts of High Energy Physics, St. Croix, USVI.

6   Elizalde, E. "A remembrance of Hendrik Casimir in the 60th anniversary of his discovery, with some basic considerations of the Casimir Effect". (2009). Journal of Physics: Conference Series. (Vol. 161, No. 1).

7   Kellermann, K. I. "The Discovery of Quasars and its Aftermath". (2014). Journal of Astronomical History and Knowledge. (Vol. 17, No. 3, 267 – 282).

8   "Exposing Black Holes Disguised in Dust". (2012). National Aeronautics and Space Administration (NASA), Jet Propulsion Laboratory (JPL). Retrieved August 2014.

9   Genzel, R.; et. al. "The Galactic Center Massive Black Hole and Nuclear Star Cluster". (2010). Reviews of Modern Physics. (Vol. 82).

10   Gobat, R.; *et. al.* "A mature cluster with X-ray emission at z = 2.07". (2011). Astronomy & Astrophysics. (Vol. 526).

11   Nayyeri, H.; *et. al.* "A Study of Massive and Evolved Galaxies at High Redshift". (2014). The Astrophysical Journal. (Vol. 794, No. 68).

12   Rahvar, S. "Cooling in the Universe". (2006). Sharif University of Technology, Institute for Studies in Theoretical Physics and Mathematics.

13   Cox, T. J.; Loeb, A. "The Collision Between the Milky Way and Andromeda". (2008). Monthly Notices of the Royal Astronomical Society. (Vol. 386, No. 1, 461 – 474).

Equations Relating Gravity and the Hubble Flow

NEWTON'S LAW OF UNIVERSAL GRAVITATION has been the foremost method of explaining the phenomenon of attraction between two massive bodies for centuries. Likewise, there is good reason for such – the equation is elegantly simple; it relates force to mass versus distance, with no other constituents -- that is, with the exception of the arbitrarily defined gravitational constant. This value, a proportionality of seemingly unknown basis, has otherwise eluded precise mathematical or theoretical definition by scientists. Even with the introduction of Einstein's theory of general relativity, and the subsequent landmark solutions which followed, this same proportionality constant is employed which traces its origins all the way back to Isaac Newton.

Henry Cavendish, an Englishman, bears the distinction of being the first known individual to experimentally measure this proportionality – albeit indirectly – via a well-known experiment which carries his name.[1] The measurement was indirect in the sense that Cavendish did not explicitly perform the experiment with such a goal to measure this constant in mind. Instead, in an attempt to measure the density of the Earth, he experimentally calculated the attraction between two heavy lead spheres. Later analysis of his work reveals that via the level of precision he exercised, derivations of the constant are possible to within about one percent of the latest, most sophisticated, modern experimental results we have today. Despite this, no one has ever been able to describe *why* this proportionality is the specific number it is with any appreciable degree of plausibility.

Perhaps, as with many things, the devil is in the details. Looking back at most of the equations in the previous section on *Hubble's Law and the Hubble Flow*, one may notice that all terms presented seem to be expressions of varying combinations of fundamental constants. We have values such as π and the speed of light, c, in proportion to H_0. In most of these expressions, we also have the constant G, which as introduced before is the *gravitational constant*, with an experimentally agreed upon value of 6.67384×10^{-11} m³ kg⁻¹ s⁻². The function of the gravitational constant G in these equations, as we have stated before – much like a great many of other equations and constants throughout physics – exist to provide proportionality. Within the equations of the Hubble flow as discussed previously, it seems to be somewhat of an oddity, given that the definitions of the other values can be clearly described. Let us consider the following determination of G:

$$G = \left(\dfrac{c}{\dfrac{1}{H_0}\pi^2}\right) {}^{kg}\!/_{m^2} = \dfrac{cH_0\,kg}{\pi^2}\,/_{m^2}$$

$$= 6.67384 \times 10^{-11}\,{}^{m^3}\!/_{kg\cdot s^2}$$

In this instance, H_0^{-1} represents *Hubble time* (expressed as the reciprocal of Hubble's constant), and c of course represents the vacuum speed of light. Since the value of c is defined as 299,792,458 m·s⁻² with no uncertainty per its exact defined *International System of Units* (SI) value, and the value of π is a fundamental mathematical constant also with no possible variability, this leaves the uncertainty of this expression entirely dependent upon the measured values of *either G, or H_0*. Knowing the value of either one of these fundamental constants with absolute certainty would permit the precise determination of the other.

The relative uncertainty of G is small – much smaller that the uncertainty of Hubble's constant.[2,3] Therefore, as a logical starting point, we should be able to substitute known values into our equation and solve for H_0 with reasonable agreement within the experimental tolerances with respect to its best known value. To better explain our derivation of H_0, We shall arrange the equation as follows:

$$6.67384 \times 10^{-11} \, \text{m}^3 \big/ {\text{kg} \cdot \text{s}^2} \times \frac{\pi^2}{c} \times \text{kg} \big/ {\text{m}^2} = H_0$$

Performing the multiplication through, we arrive at the result of $H_0 \cong 2.1971 \times 10^{-18}$ s^{-1}, which after converting units – from inverse time back to distance per time per distance – this equates to 67.802 km s^{-1} MPc^{-1}. When compared to experimental observation, this value is in remarkable agreement with the 67.8 ± 0.9 km s^{-1} MPc^{-1} figure obtained from the Planck mission data.[4]

In terms of geometry in relativity, a *geodesic* describes how a straight line in the Euclidian sense would cope with the curved shape of space-time. Such a geodesic may be characterized via the following: [5,6]

$$c^2 = c^2 \left(1 - \frac{2GM}{c^2 r}\right) \left(\frac{dt}{d\tau}\right) - \left(1 - \frac{2GM}{c^2 r}\right)^{-1} \left(\frac{dr}{d\tau}\right)^2$$
$$- r^2 \left(\frac{d\theta}{d\tau}\right)^2 - r^2 \sin^2\theta \left(\frac{d\phi}{d\tau}\right)^2$$

From this, we can reduce to a simplified expression:

$$\frac{dr}{d\tau} = \sqrt{c^2 \left[\left(\frac{E}{mc^2} \right)^2 - 1 \right] + \frac{2GM}{r}}$$

Specifically, this example comes from the line element of the Riemann manifold which describes four-dimensional space-time – the special case in the vicinity of an object with mass M.[7] Given our hypothesis that the same mechanism which causes space-time to compress around a massive object is also responsible for its dilation on cosmic scales, the required substitutions we have made should be valid in order to describe this effect.

Let us perform a through experiment using this equation. The term r specifies a distance; likewise the Hubble length is a distance we previously defined as:

$$r_{hs} = \frac{c}{H_0}$$

If we substitute r_{hs} for r, the value of this geodesic equation should be proportional to Hubble's constant when the term M equals the total mass contained within the Hubble sphere – that is proportional to the total amount of distortion caused by *all* interacting mass within the visible universe. Since the geodesic equation is in terms of $dr/d\tau$, we will set the entire equation equal to terms of H_0, over the rate of change in radius, which will simplify to units of inverse time. Let us first consider the following:

$$\frac{H_0}{dr} = \frac{dr}{d\tau} = \sqrt{c^2\left[\left(\frac{E}{mc^2}\right)^2 - 1\right] + \frac{2GM}{r}}$$

Since it follows that if the *net* total energy of our system should be zero – as would be the hypothetical case where all transient vacuum fluctuations have "cancelled" – the following should also equate to zero:

$$\frac{E}{mc^2} = \left(1 - \frac{2GM}{c^2 r}\right)$$

Accordingly, we find this is indeed the case:

$$\left(1 - \frac{2GM}{c^2 r}\right) = 1 - \frac{2cH_0}{c^2\pi^2} \times \frac{3H_0\pi}{8c} \times \frac{4\pi c^3}{3H_0^3} \times \frac{c^{-1}}{H_0^{-1}} = 0$$

From this point, we can then substitute this result back into our original equation and continue to solve, after simplification arriving at:

$$2.1971 \times 10^{-18} = \sqrt{\frac{2GM}{r} - c^2} = \sqrt{\frac{2cH_0M}{\pi^2 r} - c^2}$$

(where $H_0 \cong 2.1971 \times 10^{-18}$ s^{-1})

Solving for the term M, we find $M \approx 9.1877 \times 10^{52}$ kg, equating to almost precisely our previously calculated mass of the universe[†]. This represents strong supports for our hypothesis that the mechanisms of gravitational attraction, and the seemingly very different opposing force driving the expansion of the universe, are actually deeply related and may even be one in the same. This explanation also begins to reconcile the perceived incompatibilities between general relativity and quantum mechanics; such is a significant problem if the genesis of all mass–energy is alternatively ascribed to the process of the Big Bang.

By no means is this the first explanation of gravitation which places a repulsive force at great distances; there have been several such theories proposed in the past. It seems however, that these previous explanations fail to reconcile at *both* scales – that is, both very large distances as well as very small distances – without some *additional* proportionality factor or a reworking of the gravitation equation. One should be extraordinary skeptical of a mathematical formula which discards the elegant simplicity of *either* Einstein's or Newton's theories without giving unequivocal explanation in empirical, fundamental terms to such modifications. The proposal set forth herein seeks to avoid such.

It is often times the most essential, rudimentary, and

[†]This calculation assumes the density of the universe is very nearly or exactly the critical density, previously introduced as ρ_c. Experimental observation strongly supports that this is the case.

elegant explanations which prove to be the most correct; nature has a way about itself that tends towards the basics – that is truth is often found when we adhere to this principle of explaining things simply. When we have the ability to reduce a problem to its most elemental form, or describe things on the most fundamental level, mathematics can often be employed in ways that appear intuitive. From the explanation detailed herein – without requirements of mysterious dark matter or dark energy – it is a likely that the perceived repulsive force which drives the expansion of the universe, and the phenomenon of gravitational attraction are described by the same basic fundamental process.

1   Damour, T. "The theoretical significance of G". (1999). Institute of Advanced Scientific Studies (*Institut des Hautes Etudes Scientifiques*).

2   The Committee on Data for Science and Technology (CODATA), International Council for Science (ICSU). National Institute of Standards and Technology, United States Department of Commerce.

3   Mohr, P. J.; *et. al.* "CODATA recommended values of the fundamental physical constants". (2012). Reviews of Modern Physics. (Vol. 84).

4   Planck Collaboration. "Planck 2015 Results. XIII. Cosmological parameters.". (2016). Astronomy & Astrophysics, Manuscript.

5   Heinickle, C.; Hehl, F. W. "Schwarzschild and Kerr Solutions of Einstein's Field Equation". (2015). International Journal of Modern Physics D. (Vol. 24).

6   Collier, P. "A Most Incomprehensible Thing: Notes Towards a Very Gentle Introduction to the Mathematics of Relativity". (2014).

7   Lambourne, R. J. A. "Relativity, Gravitation, and Cosmology". (2010). Cambridge University Press. (163).

$-$ X $-$

IT'S COMPLICATED

"While physics and mathematics may tell us how the universe began, they are not much use in predicting human behavior because there are far too many equations to solve." [1]

– Stephen W. Hawking

INSOMUCH AS MERELY OBSERVING OUR SURROUNDINGS, it is easy to marvel at the endless variety not just in the vast expanses of the universe, but merely in the place where you are right now seated. Even this book, and its ideas – in the words presented, and even the materials which compose them – we ask, how could all of this exist merely by chance? We ask how is it possible, as leading cosmological theories predict, that something as complex as the human mind can arise seemingly out of nothingness – this mind which can produce such thoughts as to question itself and the origin of the universe which gave rise to it. As if it were beyond our comprehension, we may say with conviction that there was a creator of some intelligence, who divinely crafted such intricate complexities, via indecipherable layers of purposefully obscured complication; others are more skeptical and question the ideology. Surprisingly, the answer is not in the complex; it is not concealed in these layers of "complications" – but rather, it is hidden in plain sight, within the details of the most simple.

If a concept is *truly* understood, then that something *should* be easily explainable. Technical or context-specific information can be almost always be abstracted into relatable terms. In its most essential form – given at least average communication skills – one who understands something should perhaps be able to explain any concept in as few as five to fifteen words, give or take – and most certainly in a few sentences or less. Only after such summary information is conveyed, are we free to elaborate on the more complicated specifics. In denying this simple explanation, we are merely dodging the uncomfortable contention that we ourselves do not grasp the concept at hand – or such thoughts sit with unrest.

What does it mean if something is *complex*? A dictionary definition is given as *"consisting of many different and connected parts"*, but what if those composing parts are complex as well?[2] As a consequence of how the universe works – that is, the universe behaves in a consistent and predictable way – eventually we would find that all complex systems are ultimately the result of the collaboration of simple parts. We could say that a most simple part would be something like an electron, an elementary particle as known by the current extent of science. The electron exists and behaves under the most basic laws of nature, the fundamental forces of physics.

What if we wanted to instead describe what is implied by the term *complicated*? In this case, a dictionary gives us multiple contexts.[2] It seems that the definition of complicated is, well, sort of complicated. Both *complex*, and *complicated* are words which are often used interchangeably; yet often times they can mean very different things.

Anyone who has seen a drop of oil in a puddle of water can attest to the wonderful patterns of color which are produced. This phenomenon – known as *thin-film interference* – is the result of light waves self-interfering with each other as they are reflected from the surface – passing from the air, to the oil, to the water, and then back out again. Each step along the way, a ray of light is subject to a few basic rules of the universe, the same few fundamental rules which everything else in existence must obey. It is a perfect example of a simple set of rules convening to produce a complex outcome.

The vast majority of what we interact with, what we

sense and observe and see, is composed of electrons. In elementary school, we are taught that the world around us is comprised of atoms, various combinations the elements arranged on the periodic table. Anything we can touch and feel is made of some composite of the elements, much from the first few rows. However, now that we are older, we can be a bit more technical. The nucleus of the atom is very, very tiny; this is true no matter what element we choose, from hydrogen to uranium. In demonstrating such scales via analogy, if we were to represent the volume of the nucleus of a typical oxygen atom as an ordinary marble, the atom itself would be larger than a baseball stadium. While it is true that nearly all the mass is concentrated in the nucleus, the volume of the atom itself is almost completely a consequence of the electrons. As your hand grasps this book, the electrons of the paper (or digital analogue thereof) repel the electrons in your fingers and thumbs. Whenever we touch anything, we are touching the electrons bound in interaction with that substance, with the electrons of our own fingers.

Whenever light passes through a material, it mostly passes through its electrons. In doing so, an interaction occurs between the electron and the photon. Like all subatomic particles, electrons and photons both exist under the wave-particle duality paradigm; and, like any two waves in the same medium, they can interfere. Different elements can acquire different states of charge – this represents one important reason why we call them different elements in the first place. Via ionization and states of excitement, the configuration of the electrons interacting with the nucleus of the atom can adopt different shapes and configurations. This can be described, in analogy, to vibrating harmonics on a stringed instrument. Photons will interact differently with these varying configurations; light will pass through certain materials faster, while other materials it will pass through more slowly. Some materials will preferentially absorb the photons – and, differing still – other materials will reflect them away. Through one consequence, this varying interaction, or interference, is what gives rise to the refractive properties of optics.

On returning to degrees of complexity, let us examine some fundamentals of geometry. The circle is a very special shape when mathematically describing the processes and events of our physical world. The value π, which essentially describes the circular relationship, is found in more formulas than one has time or desire to count; it is completely evident that the circle *is* special. But *why* is it special? Why would any other curved shape, such as a figure-eight or a hyperbole, or a horseshoe, not hold just as much significance? Why would something which is not curved, but still is described in terms of a relationship with π, such as a triangle not captivate the attention of mathematicians so strongly? One could argue that it is because the circle is the simplest construct, both is curvature and with its absence of vertices; but then what properties convene to make it the simplest?

In the eighteenth century, a famous mathematician known as Leonhard Euler gave the world the following expression:

$$e^{i\pi} + 1 = 0$$

Deemed *Euler's Identity*, is considered by many to be one of the *"most beautiful equations in mathematics"*. The term *e* is known as *Euler's Number*, named in his honor, and describes an exponential relationship. Both terms π and *e* belong what are known as a *transcendental numbers* – a number of nature which cannot be alone expressed algebraically. The above identity is one of the methods used to prove that π is in fact not simply a useful factor of proportionality, but an inherent property of nature.[3]

One explanation of the circle – and by extension higher dimensional constructs such as the sphere – exists to state in its simplest geometric form, it is a description of something which spins. We are not speaking of the quantum mechanical variety of spin – albeit vaguely related, that is a wholly different concept; but rather, we are talking about something which actually spins, like a top. Spin in an interesting phenomenon; it is a motion which has direction, yet results in no change in position. It has the capability to introduce *inherent* asymmetry into an otherwise symmetric system, and serves as one example how it is possible for simple systems to give rise to complex systems. If we had a fully uniform cloud of hydrogen gas, in the absence of spin, nothing particularly interesting would occur. But as soon as we introduce motion of an asymmetrical nature, areas of varying density soon develop as a result of the ensuing anisotropy. Spin, whether clockwise or counter-clockwise, is a source of such anisotropy.

As an expected result, regions of relative increased density will influence their immediate vicinity within the cloud, and begin to draw in more particles, thereby increasing its mass and drawing in even greater mass, and so-on. As gravity increases, the local region of hydrogen is drawn closer and closer together; starting as a gas, then as pressure increases and thus temperature, it will transition into a plasma. In this state, positively charged atomic nuclei will electrostatically repel one another, but only to an extent past which the ever increasing gravity overcomes this. Ultimately, as pressure and temperature rise still, nuclear fusion will result. From this point, temporary stability is attained as the outward pressure of fusion prevents the cloud from further collapsing. We call this end result a *star*.[4]

Keep in mind that we have determined this is possible really only using two variables, in other words only two degrees of freedom – linear motion and angular (rotational) motion – and one force, gravity. We briefly mentioned electrostatic and nuclear forces towards the end of our star formation example. The universe we know so far is based on just a few basic rules; factoring in the other fundamental forces immensely adds to the complex situations which we observe. At the macroscopic level, the electromagnetic force gives rise to electrostatic and magnetic attraction and repulsion. On smaller scales, there is the weak nuclear force, which among other things mediates radioactive decay; and on smaller scales still, there is the strong nuclear force which governs the binding together the particles of the atomic nucleus. At all levels the force we perceive as gravity, which is the description of the shape of space and time, mediates everything.

Quite astoundingly, at some seemingly insignificantly small corner of all the vast and endless complexity of the cosmos, shortly after the birth of our planet, something which we regard as very special occurred. The "stuff" of life, *organic* molecules, this chemistry based on carbon chains, combined in patterns which were at least initially conducive to self-replication. When we look at something as complicated as a human, or an insect, or even a single-celled organism, we see this dizzying complexity which seems to defy chance. We question how all the pieces, all this machinery of organelles and ribosomes and instructions could simply fall into place – and yet the answer is surprising, they did *not*. They were in fact, *designed*, but not in the conscious sense that one may be inclined to think. Rather than the result of purposeful will by the almighty, these complicated things arose as result from the simpler manifestations of the universe and by obeying the same rules which caused these manifestations to exist.

Carbon, the sixth element on the periodic table, is a simple element containing only six protons. It is produced in stars in a few ways, but perhaps the most significant way it is produced is through something known as the *triple alpha process*. As a star continues to fuse hydrogen nuclei into helium nuclei (also known as *alpha* particles, ^4_2He), these denser alpha particles accumulate. Once critical conditions are met, chiefly after exhaustion of abundant hydrogen nuclei, they too begin to combine and fuse. Two alpha particles results in a beryllium nucleus, as ^8_4Be. The result of fusion with another alpha particles yields carbon-12 ($^{12}_6C$), a very common element – but also a very important element to the processes of life.[5]

When this carbon plasma is eventually allowed to escape the super-hot, super-dense confines of a star – a process afforded by stellar death every several million to several billion years – it may cool, and acquire electrons, and assume an atomic form. Existing as a carbon *atom*, it may now participate in electron interactions with other atoms, still all via the consequence of the same physical laws which enabled the formation of the carbon via fusion in the first place. We call the study of these electron interactions, in the range of temperature where they exist, the study of *chemistry*. The waveforms assumed in these weak interactions (due to electric attraction and repulsion) are the basis for all we know and hold dear. But their variety is no mystery – it is merely a consequence of the endless many combinations which they may assume, based on composition and proximity, and all under the same set of rules. Many of these rules are based on fundamental mathematics.

The chemistry of carbon happens to be special; but special to us only because it is important to our existence. Other elements are special in other ways; the element mercury for instance, is one of only two elements which exist as a liquid at room temperature. But what is room temperature? In the vast expanses of the universe, the notion of "room temperature" is a lonely point on a continuum which spans far beyond not only what we consider comfortable or tolerable, but far beyond our physiological limits. In certain isotopic flavors, the element of uranium demonstrates special attributes in lieu of its precarious instability, teetering on the edge of self-destruction, only if its nucleus was to be nudged by a neutron of modest speed. Such is a property we exploit for the production of electrical energy in nuclear fission reactors. Carbon is particularly special – to us – for its unique chemistry; it has the ability to combine with itself, as well as many other elements. Atomically, it is an acceptable physical size to be conducive to bonding in large, complex structures. The waveform interactions of its bound electrons resonate in the patterns which enable it to easily interact with up to four neighboring atoms – which in itself is not unexpected, tetravelency is a property of all elements in on the periodic table within the same vertical column as carbon.

In fact, it is a *periodic* feature of all elements to have a certain valency and is one of the primary reasons why elements are arranged into the "periodic" table at all.[6] It is merely a consequence that carbon is abundant, physically small, and tetravalent which make it statistically the most likely element to participate in the chemical compounds approaching the complexity of life. Yet life does not circumvent chemistry, much contrarily it is confined within and must obey, again, the same rules as anything else. The patterns we see in the organic chemistry of life, are akin to the beautiful bands of color from the drop of oil in the puddle of water; only now instead of a single ray of light interfering with itself, we have the dizzying array of electron interactions across a multitude of elements to contend with. The complexity which arises is exponentially greater; and our capacity to comprehend this vast array of simultaneous interactions is easily overwhelmed and by far simpler things.

1   Johnson, F. "Inside the Mind of Stephen Hawking". (2014).

2   Merriam-Webster's Collegiate Dictionary. Eleventh Edition. (2004). (254).

3   King, J. P. "The Art of Mathematics". (2006). Courier Corporation. (86).

4   Kennicutt, R. C., Jr.; *et. al.* "Star Formation in the Milky Way". (2012). Annual Reviews of Astronomy and Astrophysics. (Vol. 50).

5   Arnould, M.; Takahasi, K. "Nuclear Astrophysics". (1999). Reports on Progress in Physics. (Vol. 62, 395 – 464).

6   Kotz, J. C.; *et. al.* "Chemistry and Chemical Reactivity". (2012). Cengage Learning. (61).

LIFE

THE GEOLOGIC RECORD indicates that life on Earth was present *very* early in its history. The story of our planet begins approximately 4½ billion years ago, when the Earth coalesced from the debris disk orbiting a primitive sun. Geophysics and planetary science inform us of the process of accretion, via release of gravitational potential energy, results in the tremendous heating of a forming body. The early Earth was almost certainly extremely hot; for a significant period of time, the environment remained outside the range which affords the formation of complex carbon-containing organic molecules so essential to life as we know it. Furthermore, at some point in the Earth's early history, the postulated collision resulting in the formation of the Moon would have imparted enough thermal energy to momentarily result in a molten surface. Despite this, the earliest generally accepted evidence of life occurs at between 3½ and 4 billion years ago.[1]

The best empirical affirmations we have suggest that life took hold quickly, very soon after the conditions were favorable for its existence. Any sooner, and life would have to contend with the late stages of accretion, heavy bombardment from planetesimals, and molten geological resurfacing events – all of which are of course incompatible with the existence of biological organisms.[2,3] If the development of life on a planet is rare – questionably an unlikely scenario given early Earth-like chemistry and conditions – imagine how much rarer it would be for life to form at just the right point in Earth's history which maximizes its time to exist. On the contrary, hypotheses statistically point to the possibility that the organic chemistry of life is actually a *common* outcome under the conditions which allow it to exist – or at the very least not something which is statistically improbable. If this is indeed true, then it would not be a stretch to state that throughout biological evolution, even on our own planet, did not derive from a single, solitary common ancestor. Rather, it would be much more plausible to postulate that early life was a competition amongst a multitude of primordial up-risers on the initially lifeless Earth. In fact, it would be equally plausible to state that life, over its several billion year history, may have been continually influenced by sporadic episodes of "spontaneous genesis" of life-like chemistry, given the fact that it emerged so readily, so briskly. (The term "life-like" is used broadly here, at best possibly referring to unsophisticated interactions of amino acids or nucleic acids.)

If it only took this brief span of time – geologically speaking – for life to form initially, what is to say that the same process did not occur twice, given a time period thousands of times longer? What prevents this from occurring three or four times separated by even a billion years? What about even more frequently; or even just five minutes ago, somewhere – anywhere – on Earth? We do know, based on the geologic record and observations within the solar system – and to a growing extent, via the study of exoplanets – that the Earth was perhaps a most favorable place to call the cradle of life.[4] This environment long ago, just before the dawn of living things, resembled more of the crucible of an organic chemist rather than a place friendly to life as we like to commonly think of it today. It is only *after* life developed, ironically – and because of it – that we have these preconceptions of an environment which is favorable to life. (Existing in another form, life elsewhere may thrive in very different conditions.) On Earth, through its own chemistry, living organisms long ago terraformed the planet from a vigorous reaction vessel into to a state which was *no longer* conducive to the spontaneous formation of organic compounds. That is to say, that the chemistry of life though photosynthesis and cellular respiration and metabolism slowly changed the Earth from an unstable, chemically reactive environment, into the serene blue planet we know and love today.[5]

The unfortunate truth is, at this time, we may only speculate. Presently, our theories can only be tested in rudimentary ways, at best. The human mind is hopelessly inept at fathoming lengths of time even approaching our brief lifespans of a few decades – let alone spans of time millions, or hundreds of millions times longer. Even worse are we at comprehending the stupefyingly vast number of molecules at any given moment subject to the same rules, in parallel. There are many times more molecules of water in a single teardrop than there are gallons in all the oceans of the world, with each individual molecule simultaneously interacting with their numerous neighboring molecules, and the near-endless variety of substances dissolved within. Perhaps, suspended in the water, some of those substances are amino acids, peptide chains, or nucleic acids; other compounds may also be present which are aggressively reactive. Furthermore, chemical reactions on the surface are subject to the continual energetic input via ionizing solar radiation. Think of the insanely large number of interactions every *microsecond* which those molecules are exposed to. Think of the baffling number of opportunities for amino acid chains or strands of RNA to form, and to be changed by the environment. Our minds are used to the serialization of single ideas, but the universe does not work in this manner. Everything, everywhere, occurs simultaneously and under the same prescribed fundamental forces; the same set of rules.

Established life is continually in flux; the unique flexibility and changeability afforded by an instructional genetic code allows life to briskly change and adapt to its environment. Changes which promote its existence tend to stick around longer; detrimental changes conversely do not have a tendency to linger. The only required precursor is an efficient means of reproduction. RNA provides this innately; the situation with DNA is more complex, and perhaps why we see signs of its emergence later.

The study of contemporary biology supports this rapid and flexible adaptability – sometimes also referred to as "evolution" on long timescales. One instance where it can be most directly observed in the unfortunate effects of severe, acute radiation poisoning. The *sievert* (Sv) is an internationally standardized unit of measure for radiation dosage, and is equivalent of the result of absorbing one joule per kilogram of mass via exposure to some source of ionizing radiation. One sievert represents an enormous dose; on average, the level of radiation that an individual can expect to be exposed to is approximately 2.4 *thousandths* of a sievert, per year.[6]

This inescapable, incidental exposure is commonly referred to as "natural background radiation". Under these typical conditions, such natural environmental exposure poses no significant threat of adverse effects.

Even in boosting this exposure tenfold, as aircraft pilots and crew routinely experience, it still produces little evidence for clear risk. However, at levels of hundreds or thousands of times higher, even by means of a singular, momentary exposure, death inevitably ensues. When the genetic information of the cell is damaged beyond repair in devastating radiation exposure, accordingly instructions can no longer be reliably be executed. Whatever machinery within the cell which has already been synthesized – that is, existing proteins, membranes, structures – remains and functions until degraded. The process of degradation, akin to the cell machinery "wearing out", is the same process occurring in all cells at all times; the result quickly manifests itself as the organism succumbs to the effects of radiation poisoning within several hours to a few days.[7] Nucleic acids are fascinating molecules; they contain all the prerequisite information to create a human being from scratch, sustain it for the entirety of its life, and even how to reproduce itself. Our genetic libraries are unfortunately also very fragile.

This process of damage and mutation serves to illuminate just how malleable and susceptible nucleic instructions and resultant encoded proteins are to change. The mutable nature of nucleic acids represents a very important attribute; it affords the contained instructions to have the capability to change on timescales which are brief even by human standards. As long as the rate of change is tolerable (*e.g.*, low levels of radiation, environmental stressors, dilute chemical factors, *etc.*), many of these changes tend to result in nonsense instructions, and perhaps a few result in defunct, useless proteins. In the case of the latter, the unfortunate organism that receives such damaged genetic attributes usually experiences a higher rate of mortality, dependent upon the importance of utility such a gene may have conferred.

Still, statistics operates on the sheer number of genes, their small size, and ubiquitous abundance. It is likely that, throughout the vast environment on Earth, on a continual basis of some regular frequency, a useful genetic mutation occurs which somehow benefits the survival of the organism or its offspring. Such a change may activate or deactivate a gene, or change some other part of the genetic code to confer some advantage which was previously not present. This change in the genetic code is then copied over and over again, and passed down from generation to generation, until it is itself mutated, or otherwise superseded by a more effective instruction via the same process. Given the great many of copies of genetic information within the sheer number of organisms on the planet, the likelihood of a favorable change occurring in an organism at some randomized location on Earth is greater than being struck by lightning – a phenomenon which we are well aware does occur, and perhaps at a rate more frequently than one may casually expect.

Within the genetic code of most organisms today,

there are billions of base pairs – rungs on the ladder within DNA. The fact is that disruption of even a single base pair can have profound effects on the organism. The *ribosome*, which differs somewhat across life on Earth, is a curious structure – it is largely made of a few strands of nucleic acid which have folded upon themselves, augmented by a few proteins, meant to serve as a sort of efficient universal template for the rest of the genetic information within the cell. The active site of the ribosome has exposed nitrogen bases – a place for three rungs of one side of the ladder to interact – which when encountered in a certain pattern, cause the structure to distort and conform under the influence of varying strengths of atomic bonding. This creates a situation which is conducive to the binding of a single amino acid to a nearby site along the ribosome, joining one amino acid to another forming a chain as a half-strand of genetic information moves through.

There are many combinations and sequences of base pairs which correspond to specific amino acids, and some which do not; furthermore, in many organisms there are large stretches of the genetic code which exist not to code amino acid sequences at all, but rather serve to regulate its expression to varying degrees. Within genes, a single change of a base pair can completely inactivate the entire coding of a particular protein. Alternatively, a single change in a different base pair may cause the ribosome to substitute one amino acid for another – thereby resulting in a different protein with a different structure. Yet another outcome may have such a mutation result in a "halt" instruction, shortening the amino acid chain to the point of such a change, again resulting in the failed synthesis of the original protein. Outside of modifications to genes and resultant protein synthesis, varying degrees of changes in gene expression can be induced even with a single alteration of a base pair in one of these non-coding regions as in the case of *introns*, and untranslated regions.

Most mammals have the ability to produce their own ascorbic acid – otherwise known as *vitamin C* – and do so on levels several times the recommended daily intake per unit body mass as humans. Ascorbic acid is most important for collagen synthesis, but it also serves a somewhat important role as an antioxidant. Take a moment to think about our modern obsession with this nutrient; then of how almost every other four-legged furry creature on Earth produces far in excess on their own than they will ever need. Surprisingly, we humans contain the same genetic instructions to produce the enzyme required to form vitamin C; however, the copy of the gene which codes it is defective.[8] A single transposition of one base pair has made the instructions useless. This identical mutation is one small piece of evidence which points to a common ancestor, many millions of years ago and long before the divergence of the hominids, of a wide swath of primates which eventually led to humans and great apes.[†] All great apes possess this same defect, at the same location; this is remarkably outstanding given the billions of base pairs which are present in each genome. Consequently all humans, chimpanzees, bonobos, gorillas, and orangutans, in passing this defective gene on from generation to generation, must obtain all their essential vitamin C through dietary means. For most of our hunter-gather, omnivorous history, this has not posed a significant problem. It usually only becomes a problem for "modern" behaviors, including prolonged durations at sea, or in certain cases, the modern "western" diet adopted by the peoples of developed nations deficient in plant-based sources of the vitamin.

Despite the fact that a single transposition, substitution, or deletion can cause the instruction for an entire protein to become defunct, most base pairs in the genome are categorized as "non-functional", in the sense that they do not code proteins when processed by a ribosome. There occur on vast stretches of the genetic code *in between* the regions which contain the protein coding patterns. While we are just starting to discern what these regions do – previously referred to by some as "junk" code – it has become apparent that they likely serve as a regulatory mechanism for the *expression* of genes. Like the structure of the ribosome, these

---

[†] The protein *L-gulonolactone oxidase* is responsible for the synthesis of ascorbic acid from simple sugars in most plants and many animals.

stretches of DNA have the ability to conform into three-dimensional, space-filling tertiary structural shapes, and may mechanically restrict or promote access to various parts of the genetic code, even based on other conditions within the organism or the environment. And, just as a single transposition or erroneous alteration of a base pair can inactive an entire protein-coding region, it can also completely alter the assumed, three-dimensional tertiary shape of these regulatory regions. A single change, which is likely to occur any given moment, has the capability to profoundly alter the course of life. While it is true that most of these changes are corrected in a system of checks and balances within the cell -- the defect is repaired enzymatically, or cell senescence is invoked, or apoptosis is invoked and the cell self-destructs to preserve the whole organism – some of these defects are inevitably are missed. Such was likely the case in the gene which encoded the enzyme required to produce ascorbic acid, in one of humanity's ancestors long ago. Thus, if such a change occurs in the right place and at the right time, and provided it is not fatal, it becomes copied billions unto trillions of times over, and passed on to future generation after generation.

Sometimes these incidental changes – these mutations – are beneficial to the organism. Through nearly an identical change, a single base pair transposition, the gene which codes for the protein hemoglobin in sickle cell afflicted individuals causes abnormal copies of the protein (HbS) to deform red blood cells into thin, irregular, elongated shapes.[9] Having two copies of this gene – one from the mother and one from the father – is universally fatal. However, inheriting only one defective copy, along with a normal second copy, only results in a mildly anemic, but otherwise viable offspring. The abnormal shape of the blood cell has the curious side effect of making that person resistant to infection by malaria. In areas within Africa, where malaria is prevalent, significant numbers of the population carry this otherwise detrimental trait. It imparts a clear survival advantage, as the mosquito which carries the parasite bites indiscriminately. Those possessing the trait are the same group which is conferred with a survival advantage, and succumb to the parasite at a significantly reduced probability. They are biasedly selected, by their traits and their environment, as they are most fit for survival.

Modern medicine was irreparably changed with the introduction of penicillin in the early part of the twentieth century, the first widely effective antibiotic available to the masses.[10] Since this time, we have been engaged in an evolutionary arms race with some of the smallest organisms on Earth. The crux of the matter lies in the mechanism by which penicillin is effective at destroying bacteria – specifically, as a β-lactam antibiotic, it is effective through disruption of the protective cell wall via binding to certain critical proteins. Bacteria may divide as frequently as every few minutes, as opposed to years between reproductions of most animals and higher plants. As a consequence of this, over many generations slight variances develop within the genetic code, which usually either contribute no significant effect (*e.g.*, occurs on a non-protein encoding region of the genome) or a fatal outcome (*e.g.*, causes a critical protein to fail synthesis, resulting in the death or non-viability of the organism). The genetic vocabulary is composed of *codons*, which are three-base-pair-long sequences causing the ribosome to conform in which a way that is most conducive to append a specific amino acid in a concatenated sequence known as a peptide chain. The four-letter extent of the genetic "alphabet" allows sixty-four such combinations, however there are only twenty amino acids used by the vast majority of organisms. The rest of the combinations are either synonyms for another amino acids (all else unchanged, producing an identical protein), or one of the three combinations causing termination of the current protein synthesis, otherwise known as *stop codons*. The combination CAA, for instance, codes for the amino acid *glutamine*. If a mistake were to be made in copying the genetic code during cell division, which mistakenly replicated CA<u>A</u> as CA<u>G</u>, there would perhaps be no change observed in the proteins produced in subsequent generations as *both* of these codon sequences code for *glutamine*. If however, CA<u>A</u> was changed in replication to CA<u>C</u>, it would now code for *arginine*, a completely different amino acid which would undoubtedly alter the protein – a change causing the protein to conform to a different shape and therefore perform differently, or not be functional at all. In yet another case, if <u>C</u>AA was inadvertently copied as <u>T</u>AA, the protein would become "snipped" at that point in the chain of synthesis (TAA is a stop codon) and the protein produced would be shorter and vastly different in function – almost certainly nonfunctional. But, on rare occasions, the latter two scenarios produce a protein, which by chance, possesses a superior function, or performs a different function which confers to the organism more advantage than disease.

In an effort to place into scope the significance of changing just one base pair in an entire genome, take the sickle cell trait we have just discussed. The human genome contains in excess of three billion nucleotide base pairs; that is the sequence G<u>A</u>G is misrepresented as G<u>T</u>G, with billions of "letters" before, and billions of letters afterward. This minute change about three base pairs, quite literally one in a billion, causes *glutamic acid* to be substituted for the amino acid *valine* in the protein hemoglobin, the primary oxygen carrying protein in human blood. A single amino acid substitution causes a conformational change to the protein. This misshapen form of hemoglobin (through a cascade effect) causes the physical intracellular morphology of red blood cells to take on a crescent "sickled" shape. While red blood cells expressing the altered hemoglobin are at a slight disadvantage in their efficiency at transporting oxygen to vital tissues, *Plasmodium* parasites, the protozoan organisms which cause the disease malaria, transmitted by mosquitoes into the bloodstream, are unable to as effectively infect the irregular crescent shaped sickle cells laden with aggregate HbS fibril structures. This overall confers a survival advantage to the human, at the expense of slightly diminished physical stamina during strenuous activity. Malaria is a disease most prevalent in the tropics, and likewise we find that the trait is most preserved through the generations within the indigenous peoples of those regions, with greatest prevalence in western sub-Saharan Africa.[11]

It is such that occasionally, these variations cause a protein to assume a slightly modified structure, yet still retain functionality. Revisiting our previous topic, in the event which this occurs on the protein contributing to cell wall synthesis – specifically one which is disrupted by β-lactam antibiotics such as penicillin – it becomes possible for that cell and its offspring to resist the effects of the antibiotic. Through careful analysis of both the genetic code of the mutated bacteria and the proteins which it encodes, we can deduce its cause was due to a change in one or more base pairs in its genetic code.

Prevalent in popular science is the promulgation of support for the theory of evolution primarily presented as deduced from the fossil record; however, such genetic evidence is perhaps more convincing. Reasons for not promoting this as a pillar to the theory likely stem from the prerequisite scientific insight required to convincingly interpret this support. This is unfortunate, in that it causes impedance of the truth by those whom are scientifically inept. The fossil record provides great adjunct support; essentially they are photographs, snapshots of a singular moment, locked in the strata of time which validates the changes we see reflected in genetics.

Imagine a photographer taking snapshots during an infamous battle of great historical significance. Centuries later, they are rediscovered and we find that these photographs contain unique evidence which either confirms or refutes our historic understanding of the battle. First, there are the elements themselves in the photographs, that is, the people, the setting, and the objects captured. By the uniforms, regalia, and weapons used, a broad sense of time in history can be discerned. Then, there are the more subtle clues; perhaps a captured motion blur is seen; or the face of a famous war hero is recognized. Finally there is the information contained within the photographic medium itself; such includes the paper it was printed, on and the technique used to develop the image. All contribute to give us important clues toward the timing and authenticity of the photograph. Much like radioactive isotope dating analyzing the decay of uranium, potassium, argon and other elements, if the photographs are examined and determined to be from, say, an instant Polaroid camera, we have greatly narrowed the range of possible ages.

Since instant film was not invented until the 1920's, and even then commercialized decades later, we can be certain that the photographs were taken at least sometime during or after the 1920's, and most likely tens of years later. Furthermore, considering the type of film which was used, and known manufacturing dates, we can also postulate a statistical minimum age for such a photograph.

We continue with our detective work. Perhaps these photographs depict American soldiers using equipment introduced during the second world war. Yet we know that these types of cameras did not become commercialized until the late 1940's, several years after the war ended. Therefore, while we have not *completely* excluded that these photographs were of the second world war, it becomes more likely that they are of the Korean war of the 1950's. When we attempt to correlate the photographic evidence with other evidence we had collected, our conclusion is supported. Thus the additional photographic evidence reflects back onto our initial understanding, which becomes clearer.

It is this type of methodical detective work which must be applied to the fossil record in order to extract the meaningful information contained within. I feel that most people who denounce evidence obtained from the fossil record fail to understand this concept. When paleontologists and evolutionary biologists study these samples, it is *not* merely a matter of laying out similar fossils and drawing a line between two that may be somewhat "sort-of" look-a-likes. This is especially true when taken out of the context of the strata of Earth from which such fossilized remains came.

The story of our own evolution is written in each and every cell within our body – nature's own recorded account of the incremental changes over the millennia. To this end, we have just begun to scratch the surface of this vast historical repertoire over the last few decades; interpreting the fossil record as our family tree is a practice only two-hundred or so years old. In reality, this process has been accumulating and editing information with autonomy for billions of years – since the dawn of life on Earth – and if some things appear a little unclear right now, I think it is because it will take us a little while to figure out the details.

Already, we can determine that *Homo neanderthalensis*, an extinct species of hominid more commonly known as the Neanderthals, existed as a branch of the human lineage, a likely descendent of either *Homo heidelbergensis*, or else some derivative of *Homo erectus*.[12] There is substantial evidence that, *Homo neanderthalensis* partially merged back into modern man sometime before the early Euro-Asians who survived the last Ice Age. This information has been recently uncovered though the analysis of the DNA, which reveals information from the *Homo neanderthalensis* genome contributes up to several percent of the genetic composition of European and Asian peoples. Another interesting highlight of this research is that peoples of African or Aboriginal Australian descent do *not* contain this same significant amount of Neanderthal genetic information. This agrees with the hypothesis that the ancestors of modern humans originated in Africa, and later a particular group or groups migrated to Europe; furthermore, it agrees with the fact that native Africans are genetically the most diverse populations of humans today, encompassing the largest variety of ancestors from which these smaller groups migrated away from to seed the rest of the Earth.[13,14]

Another branch of the family tree which exhibits a similar history is that of the recently discovered Denisovans, a smaller and lesser known but genetically distinct subspecies predating modern hominids (or possibly a separate but closely related species to *Homo neanderthalensis*). Genetic analysis reveals that the Denisovans, who were found to exist in very small numbers in eastern Asia *circa* forty-thousand years ago, were significant progenitors of certain population groups of Pacific Islanders and the Australian Aborigines, contributing towards several percent of the genetic code of modern day humans in these regions.[15] Of course, as more evidence is gathered and appropriate analysis is performed, the story of the history of humanity will become clearer. For certain, it will be a composite; not merely our musings upon bone which we dig from the ground, but where it was found, where it fits in chronology, and what it genetic information it has to offer. Perhaps the most amazing of all is, rather than artifacts we have obtained from digging in the ground, the most convincing and thorough evidence has always been part of us – spelled out in the language of life within each and every cell of our being.

In recent years, since the initial discovery and analysis by James Watson and Francis Crick in the early 1950's, great advancements have been made towards a deeper understanding of the nucleic acids of DNA and RNA. The genetic "alphabet" is comprised of just five letters: they are adenine (A), thymine (T), guanine (G), cytosine (C), and uracil (U). C and U are synonyms in the sense that they code for proteins identically, C being found in DNA allowing the formation of a double helix, whereas U is the single-strand RNA analogue. Given the presence of these four or five constituent nitrogenous bases, they will spontaneously link together in aqueous solution to form polymer chains, *e.g.*, RNA. Consider these structures of "synonymous" components, cytosine (C) and uracil (U):

Cytosine   *Uracil*

These are relatively simple compounds, composed of a heterocyclic six-carbon ring (meaning there is one carbon atom at each unlabeled vertex of the hexagons in the above illustration) with two nitrogen substitutions, and one or two ketone groups. Carbon existing in ring structures – even aromatic hydrocarbons – is thought to be a significantly abundant throughout the cosmos, as supported by recent NASA research.[17] On Earth, we find cytosine tends to be inherently less stable, and will spontaneously react in the presence of oxygen to yield uracil, which demonstrates greater chemical stability. This describes a process which is ubiquitous in all of chemistry; the gas ozone (O_3) spontaneously reacts with itself to form ordinary, stable diatomic oxygen (O_2); hydrogen peroxide (H_2O_2) gives way to water (H_2O) and diatomic oxygen, each which are far more stable. It is the same set of rules, elements and bonding, which allows these molecules to exist in the first place. It is again this same set of rules which causes these molecules to form *spontaneously* under laboratory conditions which place the atomic constituents of the early Earth – hydrogen, oxygen, nitrogen, and carbon – in a reaction vessel with some source of energy such as electric discharge or ultraviolet light, or in some cases even intense thermal energy as found deep within the ground.

It is in fact, that nature spontaneously produces compounds which are much more complex. The mineral *abelsonite* is part of a class of mineral compounds known as *porphyrins*. These belong to a group referred to as

The complex chemical structure of the *heme* group,
as found in the biological protein hemoglobin.

"heterocyclic macrocycle compounds", which, in short, means they are composite ring compounds formed from smaller carbon rings as seen in the nucleic acids depicted opposite. An important feature of some of these compounds is the central metal atom which is trapped centrally. In humans, as well as many other animals, a similar structure gives hemoglobin its marked oxygen-carrying properties. The *heme* group is produced within the human body by proteins, which themselves are encoded by patterns of nucleic acids; this would seem fantastically remarkable, had not it been for the ability of nature to produce these porphyrin compounds spontaneously through the geochemistry of the Earth. The proteins of an organism, as enzymes, comprise the functional machinery of life; but they are merely great facilitators. Enzymes only serve to catalyze reactions which are possible on their own; the principle advantage is they do so in a greatly accelerated manner, while in close proximity to other proteins and factors – all realistic attainments which convene to yield a seemingly fantastic result.

Without a thorough scientific understanding, our intuition often deceives us. In fact, the synthesis of a simple porphyrin performed in many undergraduate university chemistry laboratories merely involves combining a chemical known as a "pyrrole" (a simple cyclic carbon compound) in the presence of propionic acid (which is a simple, common acid). Both reagents, structurally, are significantly less complex than the nucleic acids described and depicted on the preceding pages. Given the simple nature of these trivial reactions, and the commonplace occurrence of these reagents in nature, it is highly probable that these compounds, at times, existed and were freely available to fledgling life as it first developed. We can hypothesize that early analogues of cyanobacteria, which would become the progenitors of the chloroplasts of modern plants, first depended on these porphyrins in their environment, but eventually gained the ability to facilitate the external environmental reaction *internally* from their precursors, thus allowing them to exist independently of where the required compounds were locally found.

As we have previously explained, our understanding of the early Earth is such that it was likely a significantly more reactive environment than exists in the present day. Amidst these early conditions, the required elements in close proximity, and these same energy sources available – ultraviolet radiation, electrical discharge, thermal energy, *etc.* – are not specific to the early Earth, but rather to terrestrial planets of second generation stars in general throughout the cosmos. These are the conditions existing on perhaps billions of places simultaneously even if we look only in just a handful of galaxies in the local neighborhood of the Milky Way.

Recently, in the laboratory, we have been able to synthesize new letters of the genetic alphabet. This shows us that the language of the genetic code, while universal on Earth, does not necessarily have to be universal on other worlds. Within the last few years of this writing, scientists have been able to synthesize two novel complementary base pairs, artificial nucleosides known as *d5SICS*, and *dNaM*.[18] Most interestingly, when introduced into a bacterial genome, these base pairs were replicated and passed down to future generations of that genetically modified bacteria. Instead of only four letters of the genetic alphabet, there were now *six*. In other words, a hypothetical sequence of the original, natural bacterial DNA might have read something like:

ATG-AAG-ATC-GGA-CGA

Researchers modified such prototypical sequences, incorporating the above described artificial nucleosides. Representing *d5SICS* as *X*, and *dNaM* as *Y*, new sequences were constructed making slight alterations from the natural, original sequence. While likely coding for something different, such a hypothetical modified sequence might have been represented by:

ATG-A<u>X</u>G-A<u>Y</u>C-G<u>XX</u>-CG<u>Y</u>

Here, certain natural base pairs have been randomly substituted to create novel codons. What is utterly remarkable, is that no cell, to the fullest extent of our knowledge – anywhere in the history of Earth – has ever used or encountered these particular artificial, unnatural base pairs, yet the machinery of the cell is able to read and process them in their presence. This is testimony to the versatility and variety of the nucleic acid model for the replication of information.

For life elsewhere in the cosmos, if it so exists, this represents one of the countless variations its chemistry may take on. A goal for this research is such that, eventually, the modified cells would be able to utilize different and "exotic" amino acids which are beyond the collection of the twenty used by nature. Such an arrangement as described herein, would theoretically support a repertoire of tens, or even hundreds of additional amino acids. This would allow the synthesis of novel proteins with potentially unique properties and applications. It would also offer a degree of to the variety of life that might be possible in places other than our home amongst the stars.

The study of life on planet Earth is so far our best means to understanding life abroad on other worlds. While it is true that we have sent probes to other planets such as Venus and Mars; the atmosphere of Jupiter; a few asteroids; to a comet; to our planet's sole natural satellite, the Moon; and even achieved a landing the peculiar moon Titan of Saturn, the data obtained is almost infinitely scant compared to our study of life at home. At present it is unknown whether life, as we so define it, is unique to us; we are still unsure if it even unique in our solar system.

Until we can either venture to another planet and find RNA or other life-like molecules, or simulate millions of years of chemical reactions within the laboratory, we simply will not know with any appreciable degree of certainty. Life may be rare, its beautiful ensemble sung only on this meager fleck of matter and dust bound in place by gravity but for but a fleeting moment in the vastness of the cosmic abyss. Or, to our delight, a more likely scenario is we may find that the chemistry of life is an inevitable consequence of the basic rules of the universe – a product of the convening of the fundamental forces of nature – precipitating just the same as the process of fusion which powers the stars, or as do the varied chemical mechanisms which form the complex repeating structure of rocky minerals. While, on the same token, "intelligent" life (as we so define it) may very well be a rarity, given its emergence on Earth in just the twilight of its existence; somewhere on a nearby world around another star, may exist the analogue of a beautiful forest, flowers, or rolling meadows. In the case, if life is common, one may expect to find the proliferation of simple organisms, perhaps not even contained within the familiar cell membranes of Earthly life, but yet still exhibiting the capability to replicate or grow. Perhaps in its most simple manifestation we would find a sort-of "life soup", where ingredients such as polymers of nucleic acids can form long chains and copy themselves. Protein synthesis may be unique to Earth, or conversely it may be an inevitable sequela of the presence of nucleic acids; but more likely, its manifestation lies somewhere in between inevitable and frank coincidence. Those beautiful forests would be uncommon, but not implausible and certainly not impossible – yet still different in ways we do not yet have the capability to imagine.

The sun is what we would consider to be a "typical" star; even within our own galaxy there are hundreds of times more stars than there were ever years on Earth – and generations of stars which existed before them, as well as future generations which will inevitably succeed them.[19] This considers only the Milky Way; beyond, there are at least hundreds of billions of other galaxies, and proportionally a dizzying number a stars. The field of exoplanetary science has gained significant momentum in recent years, as the technology to detect extrasolar planets has advanced. Now typical are the discoveries on the order of hundreds, to thousands of new worlds each year. Many of these extrasolar planets are Earth-like in mass; planets have been discovered orbiting a wide variety of stars, some sun-like.[20] Even stars which are either smaller, or more massive than the sun – to a reasonable degree – possess habitable zones and, within, the possibility exists that they harbor rocky planets with chemistries capable of supporting the development of. Perhaps some even the ability to sustain Earth-like life. According to some recent research, possibly more than twenty percent of sun-like stars harbor Earth-like planets *within* their habitable zones.[21] These are exciting, pioneering times for extrasolar discoveries in planetary science.

To find the presence of *intelligent* life however, I would consider that to be exceedingly rare. Life on Earth has existed perfectly well for a very long time without this trait of intelligence; in fact, we are now just realizing it could very well spell the end of it, if we (the intelligent beings) are not careful. So, as we gaze upon the stars in the night sky, it is therefore life itself which I do not find implausible, out there; but given the vastness of space, and thereby the limited time frame for life abroad to evolve intelligence, I would find it quite implausible is that someone else, on some other stony, dusty crumb, is looking back at us.

1   Bergin, E. A. "Astrobiology: An Astronomer's Perspective". (2013). XVII Special Courses at the National Observatory of Rio de Janeiro.

2   Rubie, D. C. "Accretion and differentiation of the terrestrial planets with implications for the compositions of early-formed Solar System bodies and accretion of water." (2015). Icarus. (Vol. 248, 89 – 108).

3   Jacobson, S. A.; Walsh, K. J. "Earth and Terrestrial Planet Formation". (2015). Geophysical Monograph Series. (Vol. 212, 49 – 70).

4   Forget, F. "On the probability of habitable planets". (2013). International Journal of Astrobiology, Special Issue.

5   Gaidos, E.; Selsis, F. "From Protoplanets to Protolife: The Emergence and Maintenance of Life". (2006). Protostars and Planets V Conference, Hawaii. (2007). University of Arizona Press. (929).

6   "UNSCEAR 2008 Report: Volume I: Sources and Effects of Ionizing Radiation". (2008). United Nations Scientific Committee on the Effects of Atomic Radiation. (4).

7   "Acute Radiation Syndrome: A Fact Sheet for Physicians". (2005). United States Department of Health & Human Services, Centers for Disease Control (CDC).

8   Drouin, G.; et. al. "The Genetics of Vitamin C Loss in Vertebrates". (2011). Current Genomics. (Vol. 12, 371 – 378).

9   Jung, J.; et. al.; "Optical characterization of red blood cells from individuals with sickle cell trait and disease in Tanzania using quantitative phase imaging". (2016).

10   Bedoyere, G.; C. Bedoyere. "The Discovery of Penicillin". (2006). Gareth Stevens. (5).

11   Grosse, S. D.; *et. al.* "Sickle Cell Disease in Africa". (2011). American Journal of Preventative Medicine. (Vol. 41, No. 6, 398 – 405).

12   Qiu, J. "The Forgotten Continent". (2016). Nature. (Vol. 535, 218 – 220).

13   Sankararaman, S.; *et. al.* "The Date of Interbreeding between Neanderthals and Modern Humans". (2012). Public Library of Science (PLOS) Genetics. (Vol. 8).

14   Simonti, C. N.; *et. al.* "The phenotypic legacy of admixture between modern humans and Neandertals." (2016). Science. (Vol. 351, 737 – 741).

15   Racimo, F.; *et. al.* "Evidence for archaic adaptive introgression in humans". (2015). Nature Reviews: Genetics. (Vol. 16, No. 6, 359 – 371).

16   Gudipati, M. S.; *et. al.* "In-Situ Probing of Radiation-Induced Processing of Organics in Astrophysical Ice Analogs". (2012). The Astrophysical Journal Letters. (Vol. 756).

17   Lau, R. M.; *et. al.* "Evidence from SOFIA Imaging of Polycyclic Aromatic Hydrocarbon Formation Along a Recent Outflow in NGC 7027". (2016). The Astrophysical Journal. (Vol. 883, No. 1).

18   Malyshev, D. A.; *et. al.* "A semi-synthetic organism with an expanded genetic alphabet.". (2014). Nature. (509, 385 – 388).

19   Robles, J. A.; *et. al.* "A comprehensive comparison of the Sun to other stars: searching for self-selection effects". (2008). The Astrophysical Journal. (Vol. 684, 691 – 706).

20   Laughlin, G. "Exoplanetary Geophysics – An Emerging Discipline". (2015). Treatise on Geophysics, 2nd Edition.

21   Petigura, E. A. "Prevalence of Earth-size planets orbiting Sun-like stars". (2013). Proceedings of the National Academy of Sciences. (Vol. 110, No. 48, 19273 – 19278).

– XII –

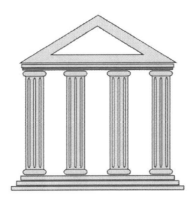

SCIENCE AND FAITH

HEREIN IS PRESENTED THESE PERTINENT ISSUES which seem to be cunningly dodged with a certain voracious tenacity all too often in these contemporary times. The seemingly infinite breadth to the topic of religious faith is broader than the day is long. The conflict of views between science and faith are, at times, entirely disparate; most problematically, no common framework exists which to reconcile differences in a logical, reasonable manner. It is amazing how far many works go to delicately skirt the issue, often taking a neutral stance as to remain indolent. Despite kind intentions, such behaviors avoid disclosing the essential truth. Sometimes the truth is thrilling, exciting; sometimes it is ghastly, frightening; nonetheless, we cannot let ourselves believe only what is comforting.

Many times, religious views are only partially based on logic – this basis also serving as a source of their appeal – and likewise require *some* evidence; yet there is often a fundamental difference in the interpretation of this evidence. A test of a religious belief, is not always exactly the same as a test of science; they are (at least at some level) incongruent, incomparable frameworks. Their respective definitions of even what constitutes evidence differs; at times, the only evidence of religion is that of its scriptures, which within, based on the attestation of man alone, are divinely revealed. However, this evidence is flimsy at best. The written word, passed on from one man to another, is mere *testimony*, and as such should be accepted only withstanding robust validation. In an analogy using law, a case brought to court on testimony only – built on hearsay – generally constitutes a weak argument. Consideration of objective evidence – which might include such entries as surveillance camera footage; or fingerprints at the crime scene; or a DNA analysis match – tends to serve much more effectively in tipping the balance.

We can state that religion reveals irrefutable evidence of itself in that it is able to predict the future – but the prophecies which do come true, whether vaguely or unambiguously, are not strongly linked to framework which validates anything more than that an author whom was clever with words. The mythology of ancient Greece may have predicted that, upon his subjects, Zeus would hurl lightning bolts from Olympus when he was displeased; yet the fact that the sky turned angry and black and lightning bolts did in fact hurl forth, contributes nothing convincing toward the presence of the displeased elder on the mountain. Similarly we can ask – what evidence is there of the Titans of *the Golden Age*, who ruled before Chronos, father of Zeus? We may continue to ask the same of any of these legends.

If the scriptures really were evidence rather than testimony, why are there such conflicting views between the varied religions of the world, all who unanimously assert their affirmation of the divine word? The works of *Lao-tzu* differ radically from the *Quran* of Islam, which in turn likewise differs equally so from the teachings of *the Buddha*. The *Bible* of Christianity purports yet another alternative history and prophecy. What of the *Vedas* of Hinduism, or the *Torah* of Judaism? More generally, what is our fascination with old books written by vaguely defined ancestors? Such is an opposing approach to the exacting and precise standards of science and reason.

Some argue that the required evidence emerges and presents itself via the fact that such testimony – the supposed divine word – on countless occasions predicts the truth through prophecy. But there is no more evidence owing that these prophecies are mere educated guesses or even coincidence without a definitive relationship. As such, this inherent vagueness of the proclamations found in religions scripture (usually without regard to any religion in particular) allow them to be adapted and interpreted to suit the predicaments of the reader. Many hold little doubt that this choice of delivery was authored purposefully. Furthermore, we can soften the meanings of words in scripture – for example, argue that something was lost in translation, that the words here now today *actually* mean something different as they were written long ago. Conversely, we can interpret something intended to be figurative as literal; we can be as inconsistent about this process of interpretation as we like, so long as our interpretation matches our ideology. Yet all the while, we concede that the words of scripture will never be in the frankness of logic. To reconcile, we may admit to ourselves that the aim of religion was never to be the cold, soulless explanations of science; but rather, it was intended to be the guide to the "good life". The distinction is all but apparent.

When the history of religion is studied, in rigor, the origins of its archaic notions over the ages becomes much more clear. Religion, as a social object, requires a society; that is, the beliefs of the collective cannot be exercised if there is no collective. We see its emergence with the first stirrings of human civilization, notably in in the region known contemporarily referred to as the Middle East. In the fertile Tigris and Euphrates river basins, extending eastward to the Mediterranean Sea, we can see the earliest archeological evidence of permanent settlements. From these small towns and villages, beginning from around 9000 B.C.E. onwards, are the first in a continuous lineage including agriculture in Sumer, the empire of Hammurabi, contributions to the rise of pre-dynastic Egypt, and following much later, the Phoenicians.[1]

Sumer, located at the heart of a geographic region known to historians as the *"cradle of civilization"*, existed in what is largely modern-day Iraq, Syria, and Turkey. These ancient peoples formed the culture which influenced the development of much of what we know as modern western society. Religion developed in pace with other social devices, namely the written word. First came the folk faiths of prehistory, who vied to the supernatural for whatever was both important and mysterious to them. The emergence of agriculture fueled the first villages, and hence there were invented gods for the rain, and gods for the grain; however, our obsession with curiosity did not leave farming misunderstood for long. Humans are both social, and ambitious – to bigger villages and establishments we sought. Resources were not infinite; as we grew with the new invention of civilization, there was competition, inevitable conflict. Our curiosity wandered; new things became important to us, and our religious allegiances changed in suit.

Through the gradual march of progress over spans of millennia, eventually the great societies of antiquity emerged. In the precedent of ancient Egypt, the great *poleis* was made infamous by the people of the *Hellenes*, who honored the gods which were relevant to their political atmosphere; correspondingly this was mirrored in the civilization of ancient Rome. It is difficult not to notice an association, a transition to progressively more structured systems of faith based on the collective political structure of a civilization's immediate surroundings. That is to say, a religion seems to be the product of the current conditions of the social-political climate; drastic changes induce paradigm shifts in religion. One such example would be the transition from dominance of Greco-Roman culture to Christianity. As the ambitions of society outgrew villages, then the *polis*, it vied towards great empires. Kingdoms require kings; likewise the dominant monotheistic systems of belief came to mirror the commander-in-chief architecture of the contemporary world. This all makes sense; crowds are often especially unruly. Under a unified system of beliefs, a religion exists as the social "glue" which provides the necessary cohesion, answers the unanswerable, and maintains order; it fosters social growth, and it quells the uncertainty which impedes culture from thriving.

Our best interpretations of history compellingly suggest that Christianity became the world's dominant monotheistic system of belief via a matter of being at the right place at the right time. Judaism, the progenitor of the Christian faith, is an ancient religion tracing its murky beginnings to the geographic epicenter of the dawn of civilization. Islam is another development of

Judaism. More than fifty percent of the world's peoples whom subscribe to a particular faith today identify with either Christianity or Islam.[2] The faiths are fervently territorial; this is paradoxical for faiths which both proclaim the "truth"; yet why does the accepted notion of such depend so much on where one is born?

In Rome, during the early common era (a period often described by historians as *late antiquity*) the popularization of Christianity approached hysteria; by around the mid-fourth century c.e., Romans were significantly split between the older classical polytheistic beliefs and emerging Christianity. The emergence of Christianity can be ascribed to the life and teachings of *"Joshua, son of Joseph"*, the man today known by many as *Jesus*. The popularization of his life and teachings form the basis of the faith. The burgeoning Roman empire and its conquests occupied Jewish lands and threatened their culture. Energizing this population (subject to imperial oppression by the Roman empire) rose a cult following of a few Jews incited a paradigm shift which was mirrored between religion and the politics of a culture.

For much of the time prior to the rule of Constantine, the relationship between Christianity and the Roman imperials was someplace between indifferent and intolerant. The radical ideas of Christians were mostly tolerated in that their small numbers posed no significant threat; however, on occasion, the empire at times exercised authority to make examples of some.

This changed amidst unrest in the third century C.E., when displeased subjects sought refuge in alternatives; Christianity, existing now for some two centuries was ready to provide such refuge. Further confounding the political unrest was a mid-third century plague which ravaged the empire. Hence, it was the salvation that Christianity promised which was increasingly seen as more attractive.[1,3]

Reaching critical mass throughout the crises of the third century C.E., both the convictions and numbers of Christians could no longer be oppressed; the Roman empire was forced to embrace its Christian culture, lest it face division because of it. The emperor Diocletian was one of the first to recognize the need for cultural tolerance; however it was famously Constantine whom initiated a state-sponsored bid to make Christianity the official religion of the imperium.

As such, the state imparted its heavy-handed influence, urging along this process to inure to its benefit. Later rulers further embraced and leveraged Christianity for various agendas – some noble, others not so much. Long a symbol of Rome and epitome of social structure, the prominence of the Senate was intentionally diminished. Further changes progressively emphasized the centrality of Christianity in the government of the state. The people of Rome, as they looked upward towards their great emperor, so did they to their almighty God who oversaw not just Rome, but all of creation. In the year 391 C.E., the emperor Theodosius by decree had moved to ban all "pagan" worship of any manner within the bounds of his empire.[4] In a matter of a decades, the theology of classical Rome fell out of vogue, albeit by forced necessity, and yielded to the new religion fit for the ambitious ruler of the people whom practiced it within his realm.

After the fall of the western empire, Christianity endured in Rome's uninterrupted successor, the Byzantine empire. When the first settlers arrived in America, the dominant native religion was displaced along with its peoples, by migrants descended from these civilizations who had forcefully imposed religious beliefs on their peoples for the gain of sociopolitical influence and power. The enduring belief in a single almighty God enabled a people to have faith in a sole, omnipotent ruler, to quiet their doubts, and unite under a common empire.

The values and customs of Byzantine Christianity had little relevance to twelfth century Native American peoples. An ocean divided the new world from the old, representing perhaps one of the best contemporary examples of geographic separation, and exemplifying the nature of religion as a sociopolitical tool. In fact, the same contrast can be observed in any two geographically isolated civilizations at any point throughout history. Generally, the more significant the geologic divide, the more radically systems of belief differ. Before widespread globalization, consider the comparison and contrast of the Americas with Europe, or with far East

Asia, or with Oceania and Aboriginal Australia. If there is anything the disparate faiths of these lands share in common, it is human imagination. The construct of religion is the fabrication of man and the culture in which he resides; the masquerade of divine intervention merely a convincing component to its solicitation. Formulating a shared system of common beliefs appears to be a practice which is an essential defining characteristic of society – much along the same lines as the development of a language, a system of writing, storytelling, and so-on.

As a testament to the nature of faith, and humanity's innate fear of change, the mirroring of religion with the radical changes in the social climate served to explain the anxieties, and curb the fears of the society. As such, it is always within this domain which religion perpetually seeks to be confined – the *unknown*. But the unknown is also the domain of science – or more accurately, the realm in which it seeks to explore. The shores of uncertainty are continually eroded by this relentless march of scientific understanding. This continuous upheaval is often unsettling, as the comforts which religion provides are threatened. Likewise, its believers become threatened, and understandably; a child will cry if it is separated from the soft, familiar security blanket. Admittedly, while the ambitions of adults are greater, our attention spans somewhat longer – and, our intellect often deeper – those emotions and feelings of the child are still within us; they constitute significant allocations of our consciousness. Sometimes, we must remind ourselves of the nature of this journey; the quest of science, is a quest of truth.

Science affords a view of the present, insomuch that it is the study of the world as it *is* – of what is most likely predicted by statistical probability, based on the events of the past. The views of religion, on the other hand, are typically more speculative; faith fosters a view which seeks to contract to the future, placing increased significance on the chance of outcome beyond what is statistically afforded by the present; it is the belief in spirituality, often aspiring to be an order of a higher power amidst the apparent chaos. Clearly, this divergence implies that there must be some cardinal differences between what science perceives as truth, and what religion perceives as truth. These differences must be quite subtle, for as basic of a conjecture, it is for certain one which is not easily explained.

But, we have already defined what *truth* is; this brings us back to the conflict between science and faith. Believing that the world (*i.e.*, the universe) will never find itself in a paradox, one accepting this definition of truth must also resolve any perceived inconsistencies encountered in the practice of their religion, their faith. If this faith becomes a socially unified collection of conjectures, then each conjecture should result in a testable hypothesis in order to even stand a chance at being accepted as truth. This presents a problem for the vast majority of religions, which often give vague and evasive explanations for deep philosophical questions.

If religion could be reduced to a fundamental question – at least popular religion in our contemporary times – I feel it would ask the question of whether or not there is a creator. However, it is unfortunately this line of thought that often times finds itself grossly overstepping certain bounds. Heated debates often ensue, and we blunderously try to assign our *feelings* without logic or scientific reason. The creator is, of course, benevolent and kind and good. It is said that he thinks, and has feelings, just like you and me. Reasoning of this kind inevitably leads to an exercise in impossible rhetoric. Those subscribing to this line of thought, are often those whom are quick to employ science in certain limited respects when it works in their favor; yet, they are just as quickly apt to dismiss the difficult and uncertain instances as something outside the realm of science. This is done arbitrarily, for no *real* reason other than it is often the most expedient way to escape an uncomfortable conversation.

This particular example is an eclectic theme. There are nearly two-hundred nations in the world at the time of this writing, not one of them having a single, identifiable religion; even amidst adversity in the presence of state-sponsored religions, significant minorities continue to exist. All told, there are more than four-thousand religions in existence on Earth in the present-day, with countless others which have fallen out of popular practice over the millennia. Regardless of the particular faith, often portrayed in a benevolent creator having a particular sort of will or desire – some motive – to act a certain way. Almost universally, this expected way of acting mimics our own behaviors, our own wants and desires. It is not this "creator" who made us in his likeness, but rather mankind which has invented the creator in the likeness of himself – akin to an orphan seeking a father figure. Towards such, there is almost universally this peculiar sense of denial, an impervious shield surrounding these beliefs; an impenetrable wall which deflects all questioning away. We are given riddles as substitutes for direct answers, and fall endlessly into traps of circular, or even self-defeating reasoning. We are told not to ask, or that it is pointless to ask because it is something we were not meant to understand; or worse, the argument is unscrupulously bridged using the mincing of words or even brute force. We are, without good reason, to believe that the intentions, workings, and feelings of the almighty creator, are beyond the realm of science, and unbounded by the laws of physics. But, that is a real problem; such implies that that the universe *does* indeed lie to us. It means, directly, that something, somewhere in the universe does not add up; it means that sometimes, one plus one now equals three, and other times it equals two. Essentially, it means that an absolute *truth* no longer exists. The continuum of reality is again threatened; such implies that eminent disaster ensues upon us in short order!

In the event that you disagree with this logic, I call upon you to give a situation where the intervention of a divine power which may be explained without breaking the laws of conservation – to give one, simple example where the energy or matter required to cause a change willed by an unworldly conscience also operates within this fundamental law of the universe. Is this an example of large scales dependent on the uncertainty introduced by the very small? If so, scale your scope down to the very smallest of dimensions, perhaps the atomic level. Conversely, does this example scale to the largest dimensions, to the volume of the visible universe?

Perhaps we can reconcile insomuch as God needs only change the trajectory of a single molecule of air in the atmosphere. As an end result of an elaborate cascade effect, his will is executed, at the same time obscuring the fact that he had anything at all to do with the situation or outcome. However, even this seemingly innocuous scenario causes calamity; that one molecule of air behaved in a way contrary to the predictable nature of the universe. Conservation of energy was violated, and this small deviation as such becomes a great concern. Even this small act would invalidate all we know and hold as true.

Yet, the argument persists, that the world is too complex for there not to be a creator. It is a question that cannot be answered with one-hundred percent certainty, whatever one's own absurd feelings or preconceptions are. Nonetheless, neither science nor the truth depend on the benevolent almighty; there exists no *valid* theory – none which adheres to the principle of *truth* – that would require a divine creator. History provides an example; we know of ancient polytheists who, long ago, believed their varied deities had made the world as it was so. In certain cultures, the constellations in the night sky were the irrefutable proof in the stories they told. The problem is, of course, that it was mankind who dreamed up these stories in the first place; the deities were an idea. In the absence of the minds of men, there were no deities, no ideas.

Science is by nature falsifiable if any one of the pillars of truth is violated – that is, either in evidence, through logic, or via mathematics. There is a certain identified safety in religion, which seeks to immunize itself to such perceived shortcomings. A fundamental difference of science is the embracing of this falsifiability – the *hypothesis* versus the *null hypothesis*. The major logical fallacy between science and religious faith presents itself in that the presence of God *cannot* be proven; conversely, many fail to realize that the only action available to science is to *disprove* that God does *not* exist.

The only scientific evidence we have towards anything at all about God are the scriptures of the various religions of the world. Some faiths lack written scripture altogether. The vague testimony of even the most well-written text provides no clear cause and effect relationship. As such, this prevents the formulation of a null hypothesis, where in the argument of a creator would seek to demonstrate there is no relationship between reality and the belief in a divine creator. There is no known exception to the rule of cause and effect. As a consequence of the former, I feel there have been a great deal of misguided theologians over the millennia, who confuse and cross-contaminate the uses of science and religion. It is a glaring contradiction to science to proclaim the existence of God as an absolute certainty; it is this immutable quality such a belief assumes which automatically negates all other possibilities regardless of evidence, cause or effect. If there is a God, then the truly scientific-minded would undoubtedly proclaim that religion must not be his creation; it is completely contrary to the predictable and testable nature of the universe. We, the scientists and scientific thinkers, have no reason to believe in the introduction of this philosophy which is pitted in *direct violation* of our definition of truth. If there is a God, we know absolutely nothing about him or her or it – to claim that we do is in irrefutable contradiction to principles of science.

Granted, I shall humbly admit that our understanding of the physical world, despite our multitude of advances, is still somewhat basic and limited. This will likely be the case for the foreseeable future. However, we only find true knowledge when we arrive at fundamental explanations, not complex conjectures. The perceived complexities of the universe do not obviously and immediately point to divine origins; jumping to such wild conclusions is very contrary to the diligent and methodical scientific approach. By this intertwining of religious ideology into scientific experimentation, we introduce bias. We ignore certain possibilities and favor others, losing the inherent objective nature that science provides. As a late astronomer and philosopher put it:

> "*Science is generated by and devoted to free inquiry: the idea that any hypothesis, no matter how strange, deserves to be considered on its merits. The suppression of uncomfortable ideas may be common in religion and politics, but it is not the path to knowledge; it has no place in the endeavor of science. We do not know in advance who will discover fundamental new insights.*" [5]

> – Carl E. Sagan

There are many devout believers who maintain that science and religion can simultaneously converge on the same truth, given that the two are somehow compatible theories. This is not to say that we cannot give examples where this is the case, but this is not *always* the case, and that is the part which science rejects. The two in fact are only partially miscible, but ultimately non-syncretizing. They are *not* two sides to the same coin, but rather wholly different paradigms.

Another logical fallacy is the irrefutable (or even merely plausible) conclusion that *anything* exists beyond or before the laws of physics, the rules which reality contracts to obey – let alone a concept as complicated and convoluted as a creator. Yet these laws are a key feature in our definition of truth; but – and we do ask – what is *their* origin? It can be said that they are merely the result of the interactions of a still even more basic set of rules – but if this is true, where did these rules come from? Such questioning goes on; unless the ultimate explanation is as elegant as a mathematical infinity, at some point we have to make the concession that *something* created *something else*; but to state that this something was God – or one of many Gods or collectively *the Gods*, or a deity – or even something purposeful like conscience thought. This is a wholly unfounded, giant leap. Scientifically, to jump to the conclusion that this "creator" (in the loosest possible sense of the word) possessed things such as intelligence or benevolence or even intent, is completely outlandish – outlandish specifically in the sense that it goes against all of the core principles of science and truth.

To say we *know* God, and of things such as his *love* and even the most remote semblance of his *plan* for us, is less than a hypothesis; no test can be conceived which validates this; it is this will or desire we have mentioned previously for God to remain forever confined to the unknown. In denial, the ridiculous notions persist, professing with utterly immovable resolve that it is *his intent* to remain perpetually and unconditionally concealed. It at best, is a wild stretch of the imagination, and at a likely worst, a delusion. Regardless of the particular denomination of faith, this evidence – the testimony – is universally written in the distant past with murky and uncertain origins (and in doing so, establishing obscurity), and refers to a calamitous scenario which occurs *soon* (thereby establishing urgency). It is a common theme, the obscurity makes refute immune to the rigors of logic, and the urgency appeals to our emotions – we are told to believe it *and to believe it soon* or forever miss the opportunity and be doomed. Like a bumptious sales pitch, the offer always stands, often times for *millennia* without an expiration date – yet the threat of urgency persists.

Even if religion could demand the same definition of truth which science demands, as we have stated earlier, it is the point where religion ceases to be, and instead identifies itself with science. It should be noted we are not saying that there is a hard and fast transition between religion and science; it is more like a graduated continuum. It is even possible for a study of a religion to *almost* be indistinguishable from the study of science. However, the separation between science and religion remains, and no matter how small it is, it always comes back to the aforementioned tendency for religion to place this increased significance on the probability of an outcome beyond what is statistically afforded by the present – assumptions towards these outcomes which lay beyond the evidence.

Generally speaking, the outcome resulting from the meddling of religion and science is usually disagreeable. Coming to mind are the likes of Socrates, Galileo, and other poor impious saviors of modern thought; and while religion in contemporary times has thankfully taken on a more benevolent persona in developed nations, a great deal of the world's violent conflicts still stem from religious incongruities.

What then – if any – is the ultimate purpose of sustaining religion? Certainly, having stood the test of time, enduring for *at least* as long the dawn of recorded history, religion must serve some vital need. It seems it is as vital in culture as food or water or shelter, perhaps even facilitating survival when the former are scarce or absent. This is because faith is an indispensable social resource, valuable not just to a leader but the individual as well; it caters to the handling of the interactions between the person and the society. This spirituality not only serves as our ethical and moral compass, it is the great mediator for the conflicted soul – and this is where I find people get confused. While consolation in faith can correct faults in emotion, it remains an ill-suited tool in the resolution of logical truth and factuality.

Humans are complex and peculiar creatures; we possess this aforementioned innate curiosity about the world. We have an inborn desire to describe the cause-and-effect relationship of nature – manifesting itself as the studies of logic and science and mathematics. But our existence is hopelessly split, the complement to this is that we have emotions and feelings; these irrational, illogical traits which somehow contribute positively to our existence in a meaningful way. It would appear as if we were left alone in the absence of emotion, we would succumb to our logical nature, seeing our continued existence as statistically unlikely; we may see attempts to prolong this existence as ultimately futile.

This is not merely a reflection of our own dizzying modern times, but all of history; the world has been quite an unforgiving place for humans and civilization to grow up. Through emotion, it is, that which gives rise to *hope*, our unwavering willingness to *try* despite the often likely scenario that we will fail. Left unorganized, emotion, feelings, and misguided hopes have the capability to wreak social havoc. It is then, the function of religion to provide the framework for which to bind together these things so that they may be used in a united way which is productive and conducive to society. We are not subscribing to the cliché that religion is man's invention intended to exert control as a means to will his own desires; no, that is a far more complicated scenario. The emotions we are discussing here are much more basic and fundamental; it is to say that religion is the social tool at the disposal of the individual.

Recall our initial observation, in which religion

contracts to the future; where religion exists as a system of beliefs which subscribes to the outcome of situations which are likely, yet still beyond what is statistically afforded by the present – this "higher power" or order. Religion is a social tool, a practice of spirituality; it is a way to cope in a world at odds against our existence. In itself, it is a *source* of hope.

So then, if the function of religion is to be a social instrument, for the intents and purposes of our objectives herein, it should be separate and exclusive. The road to spirituality, while it may mislead the mind, will never misguide the heart. From the perspective of the individual, the spiritual life is a life lived socially rich. There is this axis within consciousness, on one side we are most content with social rewards, the "social conscience"; the other, is driven by this thirst to understand, the "scientific conscience".

In fact, the word *conscience* derives its origins from the Latin word *conscientia*, which describes "inner knowledge" relating to oneself.[6] The Latin word *scientia* is simply the term for "knowledge" or "awareness", or to know something; it denotes the "external" knowledge, in contrast to *conscientia*.

If indeed religion, as our *social conscience*, reveals truth via this future-oriented optimism which we have so far described, this order of a higher power, one could reasonably deduce that the social conscience has expectations of the truth which are based on benevolent feelings. In other words, we are describing outcomes which inure to the benefit of the many – a focus of *ethics*. One can postulate that it is towards this ethical bias which the social conscience seeks to connect to, through religion, spirituality.

It may be argued that religion actually has more in common with *law* than with science. Both represent the beliefs of the collective, of society; both are social tools for the practice of social ethics. Yet no one proclaims that through law the irrefutable truth is professed; the notion is in fact ridiculous. History has taught us this on innumerable occasions – *e.g.*, the witch hunts, slave trades, and holocausts. Law is certainly not an infallible framework, nor does it masquerade as such. It is merely this aforementioned framework of ethics, which has not basis necessarily on what is true, but rather places higher emphasis on our feelings and intentions. Religion can be conceptualized in an analogous way. Overzealously, we assume that religion is infallible because it has "divine" origins; but in upholding our definition of *truth*, we must of course prove this divine origin (or disprove the alternative) to be true. Only then does religion become something greater.

The individual is soulfully opinionated. In order to function in a socially efficient manner, there must be a common agreement on pertinent ethical issues. It is therefore the function of law, and by extension the establishment of religion, which often provides this unification of the social conscience. Socially, we are tasked to create a framework whereby the objective is to converge on ethics – but how are we to determine if something is ethical? If ethics is not likely to be a yes-or-no determination, but rather a continuum, the question becomes that of ethical to *what degree*?

Ethics, based on feelings, is not a straightforward concept to define; yet, it is essential to understanding the separation which exists between the scientific conscience and the social conscience – between science and religion. Truth is something that is innately important to us, a primitive desire which elicits an emotional response. This explains how it became intertwined with ethics and religious beliefs, and why it is so fervently sought and so tenaciously defended when it is found – as well as the importance of identifying where science and religion diverge along our quest for the ultimate truth.

What of a time, when mankind has exploited science to its maximal extent? Is such a state is even possible? What of a time when the celestial mechanics of the universe are known with probable certainty? What of its origins? Consider a time when medicine has advanced to the point where we have exacting control over all biological processes, down to the molecular, even atomic level – a time when *time* is no longer so important to our waking lives, when life is no longer confined to finite spans. This describes the culmination of the practices of science, in a few disciplines, of the ideals which we aspire towards. But what if, we ask, the pinnacle of science is achieved? What are we to do with this infinite time to ponder?

This distant future (which may seem strange and perverse by the standards of our contemporary virtues) will become the mythical, proverbial utopia. Science will provide us with the tools required to reach this pinnacle of existence. Today, the realm of science is mysterious and unexplored; but just as Magellan and the circumnavigators whom have followed mapped out the world, our scientists and explorers will map out the universe. Our physicists will map out space and time; pioneers working in the vast unknown will slowly but surely advance our territory. And, just as once when humanity's awe was captivated and we became obsessed with branching out amongst the continents, we shall do the same amongst the stars. Energy, resources, time, and knowledge will all be in abundance. This is akin to the utopian description of the afterlife given by countless religions of the world since the inception of religion itself.

However, if there is one facet of religion which should irk most, it is perhaps its insistence on the acceptance of prophecy as absolute truth. What I find most upsetting is not only does this necessitate renouncement our faith in the vision of mankind, but causes us to embrace a characteristic quality of shortsightedness and laziness, rejecting logic for no other reason than it is convenient to not expend the mental energy, or experience the stress of thinking about disconcerting things.

On describing this grand future – this is *not* a prophecy; conversely it is its antithesis. Words express merely the vision of a man with a brain and a pen, whom has lived for a few short years within one tiny slice of human existence, who is able to record his thoughts onto paper. A situation which has undoubtedly been repeated billions of times before, emphasized in a select few who were obsessed with such things and gave them sufficient thought. For some reason or another, when the efforts are invisible – *e.g.*, in a manuscript written long ago, by an author whom is long dead – we tend to think those efforts never existed. The authors of every written word, I can most definitely assure you, were men and women just like you and me. There are no such things as divine prophecy. Less towards mythical figures of divinity, there should be more faith placed in the ideas capable of being generated by ordinary humans. We should take pride in the fact that we *can* have these thoughts and educated guesses about the future; our minds are powerful, insightful tools, which should not be so causally underestimated. This is not prophecy, this is not divine intervention; this is the testimony, the astute vision of man. In this probable eventuality, philosophy and ethics will become that which is most important to us. All will be provided with the clarity of ultimate understanding – or at least the knowledge of all probabilities.

1 Coffin, J.; *et. al.* "Western Civilizations: Volume 1". 17th Edition. (2011). W. W. Norton & Company.

2 "The Global Religious Landscape". (2012). Pew Research Center.

3 Fredriksen, P. "Christians in the Roman Empire in the First Three Centuries CE". (2006).   A Companion to the Roman Empire. Blackwell Publishing Ltd.

4 Tilden, P. "Religious Intolerance in the Later Roman Empire: The evidence of the Theodosian Code". (2006). University of Exeter.

5 Sagan, C. E. "Cosmos". (1980, 2011). Random House Publishing Group.

6 Cassin, B.; *et. al.* "Dictionary of Untranslatables: A Philosophical Lexicon". (2014). Princeton University Press. (175).

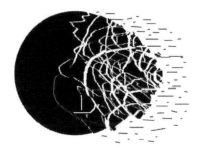

TO THE END OF THE EARTH

DESPITE ALL OUR POETIC RAMBLINGS about the infinite spans of space and time, the Earth is a place which very much has a finite beginning and an end. *Terra firma,* the solid ground on which you stand, coalesced from the ashes of a dead star which lived before the sun. Earth's position in our solar system was the consequence of multiple factors, including how much mass it accreted, combined with the initial velocity and angular momentum of this mass, numerous energy exchanges from collisions, the influence of larger planets, and other such circumstances.[1,2] It is also in an orbit which likely wandered to and fro with respect to the sun, before settling at its current place as the solar system matured. Thanks to factors such as orbital resonance, it appears as though it shall remain stable for some time to come. Venus and Mars, also residents of the so-called "habitable zone" in relation to their distance from the sun, likely shared similar geological evolutions; however Mars, being significantly smaller than the Earth and farther away from the sun, has already cooled and lost its magnetic field. The relentless solar winds, without this protective shield, have dissociated hydrogen from Mars' atmosphere and blasted it into space, and thus obliterating all free water above the surface in the process. The indications for a wet Mars in its recent geologic past – relatively speaking – are very evident, from space observations of the canyons and valleys, and old river beds; to the robotic surface observations and analysis of minerals and geologic structures which can only be formed in the presence of water.[3,4] This represents one possible fate of a planet similar to our own; this geologic cooling places time constraints on how long a world may remain habitable, in spite of a stable star. We use the term "habitable" in the sense of what is important to Earth-like life; namely water, stable climate, and some form of protection from intense solar radiation.

Planetary magnetic fields for small, rocky planets (such as exemplified by the Earth and Mars) seem to depend on the action of a liquid iron core – the *dynamo* effect as it is often described.[5] All large, planet-sized rocky bodies thus far observed seem to follow the same general composition, stratified, with the less dense mineral exterior surround the more dense iron-nickel core. Of all the common elements, iron is among the heaviest. It is produced in relatively large quantities in the death throes of a massive star, promptly decaying from ^{56}Ni with a half-life of approximately one week.[6,7] With the source of this particular nickel isotope being stellar, by the time it gets incorporated into a planet, it has largely decayed into iron. The iron sinks to the core during planetary synthesis; as intense heat comes from the conversion of potential energy, as well as radioisotope decay of other longer-lived elements the outer mineral layers thermally insulating the core. Gradually the entire planet cools, but this process, in cases such as the Earth, takes billions of years. During much of this time, the outer metallic core is hot enough to be liquid, while the inner core, although possibly hotter, is dense enough under the extreme pressure to become solid. The freely spinning solid inner core suspended in liquid iron – and via complex processes which remains with some uncertainties – is the source of the Earth's life-sustaining magnetic field. At any given time, the inner core may be spinning *slightly* faster or *slightly* slower than the rest of the planet, and this variability is theorized as one possibility contributing to the historic periodic global magnetic pole reversals recorded within the geologic record.

Planets orbiting the sun – or moons orbiting planets, by comparison – follow the very precise and predictable rules of orbital mechanics. Through this, it is quite possible to infer the mass of a planet if its orbital parameters are well known. By knowing the physical size of a planet, we can also calculate additional properties, such as its average density. To this end, combined with observational evidence, we can establish that the inner planets are rocky and each of them must exhibit heavy metallic cores – largely iron as it is the only element both abundant enough and dense enough to satisfy this requirement. Objects of planetary size are massive enough to deform under their own gravity, assuming a spherical shape; the plasticity of planetary structure on such large scales affords materials to settle into layers, or *stratify*, with the densest materials such as iron proportionately sinking towards the center.[8] This process of stratification, via the expenditure of gravitational potential energy, plausibly generates an enormous amount of energy through friction as heat; in addition to the heat produced by the accretion process itself, heat imparted due to large impact events, and radioisotope decay, the core becomes liquid, for a period of time in the history of the planet. It has been this trapped heat, so very deep underground, which has shaped and molded and even protected us, at times, from cataclysm.

Additional evidence pointing to iron metallic core formation as a typical planetary feature can be observed in the solar system's failed planet, the asteroid belt. This ring of rocky debris contains differentiated materials, much like we would find in the different layers of a planet. One particular massive asteroid *16 Psyche*, is a nearly pure metallic alloy of iron-nickel. This likely represents the differentiated core of a forming protoplanet, ultimately thwarted in some point of its evolution and torn to pieces due to the proximity of the titanic planet Jupiter and its large gravitational influence.[9]

A theoretical scenario could have been, as this protoplanet was accreting, Jupiter itself was drawing in more and more gas and dust, becoming increasingly massive; as Jupiter gained more mass, it moved closer and closer to the sun, and the infant planet beyond the orbit of Mars. At a certain point, the increasing strain from the large gravitational gradient imposed by Jupiter across the body caused it to fracture and crumble and rapidly cool. Over the eons, the larger fragments collided and progressively fragmented into the smaller pieces; some were ejected from the solar system, others swallowed up by massive Jupiter (and possibly to an extent, the other giant planets). What we observe in the asteroid belt today, are likely but a fraction of such remnants.[10] This is not completely speculative; the evidence is substantial, from what we observe today, to the constraints of orbital mechanics, and to other natural processes within the universe. Had Jupiter ended up much closer to the sun, perhaps Mars would have encountered a similar fate; conversely if Jupiter had been much less massive, there possibly could have been a fifth inner terrestrial planet.

Mars is one of the most studied planets, after the Earth of course. Though we have much to learn about its geology, we can infer that it was once much warmer and wetter in its distant past. Conversely, Venus – often regarded as our planetary twin in respect to its diameter and mass – has demonstrated quite the opposite fate. While it is possible that Venus could have been considered habitable at some point in its history, today it is an utterly hellish place. In the absence of a sufficiently strong magnetic field, both Mars and Venus have retained only high-mass compounds against the solar wind; stripping away dissociated hydrogen, and leaving behind greenhouse gasses such as carbon dioxide. Mars, being a much less massive body, has retained a only tenuous atmosphere. On the other hand, Venus, with its higher planetary mass, exhibits a dense carbon dioxide atmosphere responsible for producing a profound greenhouse effect. Its runaway effects of heating are compounded as the higher temperatures free even more trapped CO_2 which would otherwise have remained locked away in minerals. The surface temperature on Venus is several hundred degrees, independent of whether it is day or night.[11]

Yet, Venus presents another quandary; the absence of a strong magnetic field suggests that the planet's core has cooled sufficiently to solidify completely.[12] Or, another plausible explanation describes a hotter, fully fluid core of uniform temperature, lacking sufficient convection to produce a strong magnetic field. In either case, there is no evidence of a Venusian dynamo. For the former, this begs the question as to how the iron core of Venus – at almost the same size as Earth yet significantly closer to the sun – could have possibly cooled before that of our own planet. Several observations support such a hypothesis; Earth is slightly denser than Venus, by about five percent – this serves as evidence that it contains elements with heavier nuclei (*e.g.*, nickel and iron) and thus more likely to have retained more primordial heat.[13] However, perhaps the single most important answer may be hidden in the most prominent object in the night sky – *the Moon.* It is theorized that, early in the formation process of the Earth, another large protoplanetary body (in prevailing theories, given the name *Theia*) collided with our own infant planet, ejecting a colossal amount of material which eventually recoalesced into the Earth–Moon system.[14] This process, aptly named the *giant impact hypothesis*, would have injected huge amounts of energy, in the form of heat. (Even alternative hypotheses which call for multiple smaller impactors as progenitors of the Moon would have had the same heating effect.) Such vital heat has sustained our magnetic field for much longer than that of Venus or Mars. Furthermore, as evidenced by the disproportionately small heavy iron core of the Moon, the Earth retained a higher portion of its iron core, surrendering only the lighter upper layers of the Earth for the formation of the Moon.[15] This is likely a contributing factor to the aforementioned density disparity with respect to Venus. Year after year, the core radiatively cools, the outer liquid layer depositing onto the inner solid mass. Once the core completely solidifies (*i.e.* freezes) the dynamically generated magnetic field can no longer be sustained, and unimpeded, the solar wind is free to engage the atmosphere.

The alternative, plausible hypothesis for a liquid Venusian core can also be argued; such implies that the interior of Venus has *not* cooled faster than the Earth, but rather it has remained much hotter. Venus is the only major planet in the solar system observed with a retrograde rotation; it also has apparently undergone a global resurfacing event within the last several hundred million years. The latter implies that recently – in astronomical terms – the entire surface of Venus was molten. Venus contains more volcanos on its surface than any other known celestial body – many thousands – yet most perplexingly none are active, suggesting a recent, but short-lived cataclysmic event. Together, these observations are compatible with Venus recently merging with a massive, retrograde natural satellite. Such a satellite would necessarily exist in a decaying, unstable orbit; its absorption with Venus would also likely result in large decrease in angular momentum, and tremendous heating. Both theories would explain the recent geological resurfacing, as well as the slow retrograde rotation of the planet. In any case, the disparity between these opposing theories certainly makes Venus an interesting target for conducting future science. Both of these scenarios could also represent possibilities for the demise of a potentially habitable Earth-like planet.

In terms of biology, life has flourished on our planet for a long time; the study of archeobiology reveals that life has existed, at least in its most rudimentary single-celled form, for roughly 3½ billion years. Given the fact that this was likely prolonged by the thermal energy imparted by the hypothesized Theia impact, we can consider ourselves fortunate for a planet of our size. Within the confines a planet's existence, the habitable time for life is finite and a small part of this. On Earth, aside from the eventual failure of the protective magnetic field, we have bigger problems. Humans are innovative, and we could likely use technology to survive in the absence of a protective magnetic field – and possibly even use the solar wind to our benefit. However, there is one fate of the planet that is inescapable. Our sun, which has been this great giver of life, is also our shining angel of death. Year after year, as it consumes its hydrogen fuel, transmuting it into heavier elements, and little by little, it becomes hotter, and more luminous. It is nearing the end of its lifetime as a main sequence star, and as it transitions into its next phase of stellar life, it shall consume the Earth, our beloved home, entirely.[16]

Informally in conversations of popular science, one may occasionally hear the casual gossip portraying the end of the sun, reconciling it as some event happening so far in the distant future that humanity is completely removed. Most commonly described is that the sun, as a middle-aged star, will somehow "burn out" in four or five – or even more – billion years from now. I feel as though this view regards this process as benign, a sort-of gradual dimming of the sun until it looks indistinguishable from the other stars in the sky, the Earth sinking into a quiet, tranquil chill. This could be nothing farther from the truth. The stark reality is – astronomically speaking – our planet as we know it and cherish it, is not long for this world.

Anthropologists can deduce that mankind, in one form or another, has roamed the planet for approximately two million years. While such time is considerable, we can all agree in a geologic sense, this is just really a brief span of time. One of the more recent great creatures of dominance on the planet were the dinosaurs, a position they held for more than 135 million years.[17] The ubiquitous ant, an insect which has endured for a similar length of time, has survived great extinction events and continues to thrive to this day.[18] The oceans, possibly as some believe the progenitor of all life, have been host to fish for somewhere around half a *billion* years; with certain species remaining essentially unchanged for stretches spanning hundreds of millions of years.[19] Despite this, and these amazing spans of time, the breadth of history in which life has thrived and flourished in all its endless variety, is now in its twilight. We are actively experiencing what is close to mother nature's ultimate creation, the best and most refined revision. The sun, which is a vital life-giving resource, that which puts nearly all the energy into our ecosystems to sustain them, will soon become a threat.

Each and every day, the sun is steadily getting bigger, hotter, and brighter. The solar output is gradually increasing at an accelerating rate. Since about the dawn of life, the sun has provided a stable 3½ billion years of energy output from the nuclear reactions it sustains. This period of stability however, is now coming to an end. We can witness this process occurring in the near-infinite plethora of other sun-like stars in the sky. The question begging to be asked then becomes that of how long before the fall of life on Earth.

The astronomical study of these stars witch are similar to the sun has greatly constrained the possibilities. Our current best models indicate that the answer lies somewhere between five-hundred million, and one billion years from the present. One may find this a shocking revelation, in the sense that the Earth's life-sustaining years are coming to a close. Somewhere between seventy-five and ninety percent of the Earth-habitable years around our star are already gone. Accordingly, relatively soon in the cosmic sense, the increased luminosity of the sun will disrupt the geochemical cycles of the planet such that life will no longer be sustainable. Eventually, the oceans will evaporate, disappear, and the temperature of the planet will soar. This sun-seared Earth will persist for several billion years in this state, gradually becoming ever more scorched, although life would have been eradicated very early on in this process.

The sun, our once brilliant symbol of hope, will become a harbinger of doom, as it continues its relentless expansion.[20] When it reaches approximately twice its current age, the sun will have transitioned into a *red giant* – smoldering with an angry devilish glow, possessing a radius encompassing roughly twice that of the present-day orbit of Earth. Through this process, the sun will exhibit significant material ejections; as a consequence of this mass reduction, the orbits of the planets will be proportionately pushed outward. Despite this, it is very likely that the Earth will still be inevitably succumb, consumed and torn apart by the star which made it such an interesting place for so long.[21,22] Save for what we have launched forth into the cosmos to never return home, all evidence of our existence will be obliterated. Gone will be our terrestrial culture, and all evidence thereof; erased will be the entire fossil record of a rich and diverse planet. Everything of the Earth is destined to be dramatically destroyed, as it crumbles into the fiery convections of a dying sun; left only are we with the bittersweet thought that perhaps one day it shall be released back into the stellar medium again for the formation of a new star and a new generation of planets. Such is the tragic, yet beautiful cycle of stellar life and death. This place – the Earth – our home, shall be, from ashes to ashes, dust and so forth, for infinity into the future.

1   Desch, S. J. "Mass Distribution and Planet Formation in the Solar Nebula". (2007). The Astrophysical Journal. (Vol. 671, No. 1, 878 – 893).

2   Batygin, K.; Laughlin, G. "Jupiter's decisive role in the inner Solar System's early evolution". (2015). Proceedings of the National Academy of Sciences of the United States of America. (Vol. 112, No. 14, 4214 – 4217).

3   Wordsworth, R. D. "The Climate of Early Mars". (2016). Annual Review of Earth and Planetary Sciences 2016. (Vol. 44, 1 – 31).

4   Vincendon, M.; et. al. "Near-tropical subsurface ice on Mars". (2011). Geophysical Research Letters.

5   Dormy, E.; et. al. "Mechanisms of Planetary and Stellar Dynamos". (2013). Proceedings of the International Astronomical Union Symposium. (No. 294).

6   Truran, J. W. "^{56}Ni, Explosive Nucleosynthesis, and SNe Ia Diversity". (2011). Journal of Physics: Conference Series. (Vol. 337, No. 1).

7   Nadyozhin, D. K. "The properties of Ni $\rightarrow$ Co $\rightarrow$ Fe Decay". (1994). The Astrophysical Journal Supplement Series. (Vol. 92, 527 – 531).

8   Russel, D. G. "Solar and Extrasolar Planet Taxonomy". (2013).

9   Sarid, G. "Erosive Hit-and-Run Impact Events: Debris Unbound". (2015). Proceedings of the International Astronomical Union Symposium. (No. 318).

10   Minton, D. A.; Malhotra, R. "A record of planet migration in the Main Asteroid Belt.". (2009). Nature. (Vol 457).

11   Bengtsson, L.; *et. al.* "Towards Understanding the Climate of Venus". (2012). Springer Science & Business Media.

12   Baraffe, I.; *et. al.* "Planetary internal structures". (2014). Protostars and Planets IV, University of Arizona Press.

13   Williams, D. R. "Venus Fact Sheet". (n.d.). National Aeronautics and Space Administration, Goddard Space Flight Center.

14   Quarles, B. L.; Lissauer, J. J. "Dynamical Evolution of the Earth–Moon Progenitors – Whence Theia?" (2014). Icarus. (Vol. 248, 318 – 339).

15   Martel, L. M. V. "The Moon at its Core". (1999). Hawai'i Institute of Geophysics and Planetology.

16   Schröder, K.-P.; Smith, R. C. "Distant future of the Sun and Earth revisited". (2008). Monthly Notices of the Royal Astronomical Society. (Vol. 386, No. 1, 155 – 163).

17   Olsen, P. E.; *et. al.* "Ascent of Dinosaurs Linked to an Iridium Anomaly at the Triassic–Jurassic Boundary". (2002). Science. (Vol. 296, No. 5571, 1305 – 1307).

18   Schultz, T. R. "In search of ant ancestors". (2000). Proceedings of the National Academy of Sciences. (Vol. 97, No. 26, 14028 – 14029).

19   Nelson, G. "Origin and diversification of teleostean fishes". (2006). Annals of the New York Academy of Sciences. (Vol. 167, No. 1, 18 – 30).

20   Kalirai, J. "New light on our Sun's fate". (February 2014). Astronomy Magazine.

21   Talcott, R. "Earth's deadly future". (July 2007). Astronomy Magazine.

22   Ramirez, R. M.; Kaltenegger, L. "Habitable Zones of Post-Main Sequence Stars". (2016). The Astrophysical Journal. (Vol. 823, No. 1).

EPILOGUE

THE YEAR WAS 1912, the brisk sea-spray on a chilly April morning stung the face of a westbound and wayward traveler, as he marveled at the pinnacle of human innovation. The ship, the newly christened *Titanic*, was the largest, and one of the fastest modes of long distance travel. The flighty airplane, in its infancy, was still largely considered a plaything of wealthy eccentrics and thrill-seekers; a time for practical travel for the masses was still decades away. Unfortunately – and infamously – disaster struck, and just a few days later the *Titanic* sank into the cold abyssal depths of the North Atlantic; but the era it helped usher in sailed on.[1] A mere forty-nine years to the day of the cold brisk morning when that hopeful journeyman was struck in awe of mankind's then great accomplishment, the Soviet pilot Yuri Gagarin was hurtling around the planet almost incomprehensibly faster aboard *Vostok 1*, as the first man to orbit the Earth.[2]

When we look back upon the implications for first understanding a fundamental force, electromagnetism comes to mind. Shortly after discovering *what* it was, we began to ascertain its nature, then control it; the world was never again the same. This is the motivation for pursing these analyses, even in the cases where our ideas do not converge. All we have to do in order to justify such effort, is to imagine the implications of not just understanding the curvature of space-time, but being able to control it. What wonders did the pioneers of electricity and magnetism in the eighteenth and nineteenth centuries foresee was possible? How many of their contemporaries thought of it merely as a frivolous curiosity?

If there is one thing that always seems to be revolutionized on scales which seem to defy even science fiction, it would be *travel*. Two-hundred years ago, no invention of man could transport him to and fro faster than legs and feet. The horse represented the state-of-the-art means for moving from point A to point B. Then, mankind discovered new, concentrated forms of energy – fossil fuels – which propelled us into first the industrial revolution, and then a technological one. We could fire our furnaces with the then seemingly limitless coal to produce steel on a grand scale to build things bigger and stronger; we had the fuel to power steam locomotives, and faster we went. We discovered oil and gasoline, and then even more potent forms of energy, and propelled ourselves faster still. And all the while this was happening, we were learning about this new force – electromagnetism. With this technology, we could produce energy in one place, and transport it and use it in another; we could concentrate it, or distribute it. Our capabilities rapidly continued to grow. We invented the airplane and powered flight, and the world shrunk; then we invented the spaceship and shrunk the world even smaller – just a blue marble outside the window. Enter the deep space probes of the *Voyager* program and the world in its entirety became nothing more than a *"pale blue dot"*, as famed astronomer Carl Sagan once described.[3,4] In fact, it was in fact none other than he who requested of NASA to command the *Voyager 1* space probe to point its camera towards the Earth, from beyond the orbit of Neptune at the edge of the solar system, to capture this most famous portrait of our home. It is clear that he possessed the foresight to recognize such an auspicious moment in our history on which to reflect. Such moments are occurring more frequently, these days.

What will the next great paradigm in travel afford us to do? Imagine the wonder and awe felt by those people in the Apollo era, seeing the entire globe of the Earth taken in a single photograph by a space-faring astronaut; or maybe you are one of them and can recall first hand, for it was not terribly long ago.[5] In fact, there are some still among us who can recall mankind's very first spaceflight; or man's first transatlantic flight; or, but for a very few, whom can recollect the time of that traveler's cold brisk April morning. In a span of no more than one-hundred years, we have gone from being bound to the Earth in carriages, to riding chariots on the Moon, to photographing the edge of space and time from great observatories orbiting high above; all of these were products of the discoveries and understanding of electromagnetism mastered in the century prior. We have pressed on to sail ships through the solar system, and beyond. What awe inspiring achievements will be made in the next hundred years from a new century of research? How far, and how fast shall we go? What wonders will we be able to see?

Perhaps one day some descendent of ours will visit the Earth for the first time and become homesick of Mars; or maybe blissfully reminisce of a Galilean moon. Or further still, will be those who will never know how to long to feel the warmth of the sun upon their face. There will inevitably be a time when a human-carrying spacecraft will leave our world never to return again; aboard, its travelers committed to venture on into the void forever.

All things considered, I find the fact that it has even been possible for us to escape our planet at all, to be utterly amazing. It would seem the odds of doing such are so largely not in our favor, but at the same time, it is what makes us unique from all other life on Earth – and thus far we have seen, anywhere else. Our culture, our civilization, has the capability to endure for eons; but if we are to do so, we must learn to better appreciate our brethren, and the place we call home, wherever that may be. Eventually, we must spread out amongst the stars. Us humans, with our infinite reserves of hope, have aspirations to endure forever; but forever is a long time, and the world is a fragile place.

1   Beesley, L. "Loss of the SS Titanic". (2002).

2   "United Nations Human Space Technology Initiative (HSTI)". (July 2013). International Academy of Astronautics, 8th IAA Symposium on the Future of Space.

3   "Voyager: The Interstellar Mission". (n.d.). National Aeronautics and Space Administration (NASA), Jet Propulsion Laboratory (JPL). Retrieved January 2017.

4   Sagan, C. E.; Druyan, A. "Pale Blue Dot: A Vision of the Human Future in Space". (1994). Random House Publishing Group.

5   Petsko, G. A. "The blue marble". (2011). Genome Biology. (Vol. 12).

Final Remarks and
Special Thanks

Thank-you again for purchasing this book. I hope you have found this title to be an enjoyable and engaging journey on exploring and understanding the complexities of the universe, and that the reading was informative, entertaining, and worthwhile. Like most things in life, this work has been made possible not merely by the effort of one person, but of many. Special thanks goes to my close friends and colleagues, who have helped shape this book in too many numerous subtle little ways to possibly recollect each of them; and to my family, who has been supportive, receptive, and helpful along the way, and for always being encouraging in general with respect to my various ventures; and to everyone else over the years, to the innumerable others who have inspired me to write in ways both large and small, who have helped make this book a reality.

Deep Stuff

Made in the USA
Middletown, DE
20 July 2023

35492798R00168